CIO Leadership for Public Safety Communications:
Emerging Trends and Practices

DR. ALAN R. SHARK

Executive Editor and Contributing Author

With a special foreword by Zachary Tumin and William Bratton, joint authors of the book, *Collaborate or Perish—Reaching Across Boundaries in a Networked World*

PUBLIC TECHNOLOGY INSTITUTE | ALEXANDRIA, VIRGINIA

Copyright © 2012 Public Technology Institute
All rights reserved
Printed in the United States of America

No part of this publication may be reproduced, stored in, or introduced into a retrieval system, or transmitted, in any form, or by any means (electronic, mechanical, photocopying, recording, or otherwise), without the prior permission of the publisher. Requests for permission should be directed to info@pti.org.

Limit of liability/Disclaimer of Warranty: While the publisher and author/editor have used their best efforts in preparing this book, they make no representations or warranties with respect to the accuracy or completeness of the contents of this book and specifically disclaim any implied warranties of merchantability or fitness for a particular purpose. Any advice or strategies contained herein may not be suitable for your situation. You should consult with a professional where and when appropriate.

Library of Congress Cataloging-Publication-Data

Shark, Alan
CIO Leadership for Public Safety Communications—Emerging Trends and Practices
p. cm.

ISBN-13: 978-01478304715
ISBN-10: 1478304715

1. Chief Information Officers.   2. Information technology-Management.   3. Leadership.   4. Public Safety.   I.Title

Public Technology Institute
1420 Prince Street, Suite 2
Alexandria Virginia 22314
www.pti.org

PUBLIC TECHNOLOGY INSTITUTE PUBLICATIONS

Seven Trends That Will Transform Local Government Through Technology (2012)

Web 2.0 Civic Media in Action—Emerging Trends & Practices (2011)

CIO Leadership for State Governments—Emerging Trends & Practices (2011)

Shared Services and Municipal Consolidation—A Critical Analysis (2011)

Local Government Energy Assurance Guidelines–Version 2.0 (2011)

Beyond e-Government—Measuring Performance (2010)

e-Health—A Global Perspective (2010)

Digital Governance in Municipalities Worldwide—A Longitudinal Assessment of Municipal Websites Throughout the World (2010)

CIO Leadership for Cities & Counties—Emerging Trends & Practices (2009)

Local Government Energy Assurance Guidelines (2009)

Beyond e-Government & e-Democracy—A Global Perspective (2008)

Contents

Preface ix

Introduction xiii

Forward xvii
By Zachary Tumin and William Bratton

Acknowledgements xxi

Chapter 1: Policing in the World of Social Media—A Guided Tour with @PhillyPolice . . . 1
By Karima Zedan, Charles H. Ramsey, Nola Joyce

Chapter 2: Public Safety Enters the Collective Intelligence Era 11
By Paul Steinberg

Chapter 3: Successful Technology Development in Public Safety 27
By Leonard Scott

Chapter 4: Public Safety IT Budgeting 33
By Jeff Webster

Chapter 5: Collaboration Technology in Public Safety Communications 43
By Paul Wormeli

Chapter 6: Public Safety Wireless Communications—Past, Current, and Future . . . 51
By Chief Harlin R. McEwen

Chapter 7: Concepts on Information Sharing and Interoperability 57
By John M. Contestabile

Chapter 8: If You See Something, Send Something—Public Safety and the
New Media 75
By Dr. Alan R. Shark

Chapter 9: IT Governance in Public Safety 87
By David Whicker

Chapter 10: Innovation in the Homeland Security Enterprise—Ensuring New
Technologies Meet End User Needs 99
By Brad Pantuck

Chapter 11: The Nationwide Public Safety Wireless Broadband Network...... 107
By Bill Schrier

Chapter 12: Sensory Overload Considerations for Next Generation 9-1-1 Systems ... 119
By Ronald P. Timmons

Chapter 13: Geospatial Information Systems (GIS) and Their Evolving Role In Emergency Management and Disaster Response............................. 135
By Alan R. Shark, in consultation with members of the New York GIS community

Chapter 14: Spatial Intelligence in Public Safety 149
By Paul Christin

Chapter 15: Cyber Security and Information Assurance for Long Term Evolution (LTE) 155
By Greg Harris

Chapter 16: Apps for Government on the Nationwide Public Safety Wireless Broadband Network... 165
By Bill Schrier

Chapter 17: Critical Operations & Infrastructure 177
By Jeremy Smith, Lori Kleckner, Sherri Powell, Louisa King, Sharon Counterman

Chapter 18: A Perfect Storm: Managing Court Technology in the 21st Century... 189
By Tom Clarke

Chapter 19: Assuring Change Management Success During Technology Project Deployments 197
By Ronald P. Timmons

Chapter 20: The Power of Video to Save Lives........................ 213
By Bob Stanberry and Jennifer Bremer

Chapter 21: Integrated Risk Management—A Public Safety Perspective 221
By Mike Kenel

Chapter 22: A Regional Approach to Public Safety Communications and GIS... 237
By David Shuey, Christine Grimmelsman, Ed Dadosky

Chapter 23: Continuity of Operations (COOP) Automation Program 249
By Robert Jones

Chapter 24: Strategic Procurement and Public Safety Communications—
Managing Risk and Exploiting Opportunities 263
By Arthur S. Katz

Chapter 25: The Story of Technology, Integration, and Collaboration to Support
Fire Incident Response and Command 279
By Dan Rainey and Jason McKinley

Chapter 26: Integrated Public Safety Operational Control Centers. 291
By Daniel Proctor and David Kipp

Chapter 27: Interoperable Communications in Emergency Management 309
By Everett L. Davis

Chapter 28: Critical Components for Survivable, Sustainable, and Flexible
Communication Centers. 319
By Ian Reeves

Preface

On behalf of the leadership of the Association of Public Safety Communications Officials, International (APCO) it is with a great deal of excitement that I present to you *CIO Leadership for Public Safety Communications: Emerging Trends and Practices*.

This book was written with two goals in mind:

- to provide APCO members, and those beyond our vast network of professionals, with real-life examples of exciting and varied "snapshots" of how technology "thought leadership" and the drive for innovation is currently impacting the public safety landscape; and

- to educate our local officials—executives, elected leaders, and others interested in public safety communications—on the leading practices, trends, and opportunities that today's technologies can offer in the field of public safety communications.

As you read through the following 28 chapters, I think you will agree that we have accomplished our goals!

The development and publication of *CIO Leadership for Public Safety Communications: Emerging Trends and Practices* was funded by a grant from the Public Safety Foundation of America (PSFA). The PSFA was established in January 2002 by APCO with the objective to provide critical funding and technical support to public safety answering points and local emergency response officials, and to promote cooperation among public and private groups to support the public safety communications community.

I am particularly pleased that this book was developed for APCO and the larger public safety community by the Public Technology Institute. Created by and for city and county governments to address the technology issues and opportunities impacting local government, PTI and APCO have a close and excellent working relationship. PTI and APCO have worked together to produce webcasts, conference educational sessions, white papers, and now this book. I would like in particular to thank Dr. Alan Shark, Executive Director of PTI, for serving as executive editor of this book.

Many APCO corporate partners contributed to this book, and I thank them for their continued commitment to both APCO International and the public safety community.

I hope that as you read through this book and learn from the many wonderful examples and thoughtful insight that each chapter's author has shared that you take time to consider how what you are experiencing through these pages can be applied to your community and your job as a public safety professional.

Sincerely,
Gregory T. Riddle, RPL
President, APCO International

About APCO International

The Association of Public-Safety Communications Officials, International (APCO) is the world's largest organization of public safety communications professionals. It serves the needs of public safety communications practitioners worldwide—and the welfare of the general public as a whole—by providing complete expertise, professional development, technical assistance, advocacy, and outreach.

APCO International commits to strengthen communities by empowering and educating public safety communications professionals. For more information, visit the APCO website at www.apcointl.org.

About The Public Safety Foundation Association (PSFA)

The PSFA, a 501(c)(3) charitable organization, was established in January 2002 by the Association of Public-Safety Communications Officials International (APCO). The PSFA's objective is to provide critical funding and technical support to public safety answering points (PSAPs) and local emergency response officials.

Funding for the PSFA has been provided by a variety of sources, including donations from corporations, APCO members and staff, and the Wireless E-911: PSAP Readiness Fund, a non-profit organization established by Nextel Communications and dedicated to supporting the timely implementation of wireless E-911. For more information, visit the PSFA website at psfa.us.

About PTI

Public Technology Institute (PTI) serves as the national voice for technology leading practices and technology thought-leadership within local government. A non-profit, 501(c)(3) organization, PTI offers online training, publications and books, and conferences designed for the local government technologist; creates partnerships between government, private industry, and federal agencies to ensure that cities and counties have access to the latest and most effective technology solutions; and partners with national media and academic institutions to showcase local government technology programs and practices.

CIO Leadership for Public Safety Communications: Emerging Trends and Practices is the latest book in a series that PTI initiated more than two years ago to encourage technology thought leadership within government. Other books include *CIO Leadership for Cities and Counties: Emerging Trends and Practices* and *CIO Leadership for State Governments* developed in collaboration with the National Association of State CIOs (NASCIO).

For more information, visit the PTI website at www.pti.org.

Introduction

Public safety communications in contemporary society is no longer about two-way radios. Today, we have unified communication systems embracing voice, data, and video in real-time. Since September 11, 2001, public safety communications have been revamped at every level of government. Moreover, while technology continues to provide newer and better tools in every sector, the issue of technology leadership remains the most critical ingredient. Without effective leadership, even the best tools will fail us.

CIO Leadership for Public Safety Communications: Emerging Trends and Practices is part of a series of thought-leadership books published by the Public Technology Institute (PTI) that focus on the latest technology trends and practices. The title of this book may sound rather strange—mainly because few have the title of "chief information officer" in the public safety arena, yet just about every local police force and fire agency has someone, part-time or full-time, who is their go-to person for information technology. Cities and counties across America are finding that the long-established silos that exist between public safety and the main city/county technology infrastructures require a new look and, hence, new forms of collaboration and effective governance. Relationships between such silos can be a cause for friction and can lead to inefficiencies and mistrust. Managing, collecting, distilling, analyzing, and storing information is critical. Today, we can cull from multiple databases, unstructured and "big data," social media data, and emails, which categorically require vast resources to make sense of on a real-time basis. Greater leadership, innovation, vision, and agility are the new mantras.

As technology advances and demands on government evolve, the need for agile systems will continue to grow. In light of recent rapidly advancing commercial technologies, combined with rapidly changing community challenges, it appears that all local government leaders are struggling to keep pace, manage expectations, and provide services that meet the needs of the citizenry. Increased security threats from terrorism, gang/mob activities, drug-related problems, school and domestic violence, and even changes in weather patterns have made communities more reliant on local government. Residents are becoming more demanding with increasing expectations on their public officials to properly prepare, respond, and serve during emergencies even though the decline in budgets has reduced the number of government employees in critical fields.

These new challenges require a completely new level of agility that is not present in most local governments today. Many local governments, despite their best efforts, are simply not prepared to provide agile services that are similar to the agile services provided by commercial businesses and even social associations.

Sharing and active collaboration are the key to agile local governments. Sharing ideas, models, and even technology itself will begin to open doors to agile acquisition, delivery, and transition of future capabilities, resulting in continuous improvement in all public services. Creating agile procurement at a local government is an outstanding first step, but it is just a start. In our proposed model, we also believe that stakeholders at all levels should be brought into the development cycle as early as possible. Local, state, and federal representatives must work together on a designated project (with well-defined and limited scope) during the development process. End-user participation is essential to rapidly develop and prototype platforms that can be modular and easily tailored to fit the unique requirements of the end-user community. Projects should include incremental field trials starting with a small/limited number of users, where user feedback drives rapid improvement of the technology through iterative modification/trial cycles.

End users such as public officials (police, fire fighters, medics, residents, etc.) are turning more and more toward user input. This is not only beneficial, it is essential. Extensive participation is ideal, and the amount of participation is directly proportional to the success of any service offering in the cloud. Many local governments already appreciate the potential value proposition associated with migrating to a combination of private, public, and hybrid clouds to support a variety of community services as well as enhance community interaction/communications with citizens. Perhaps the most exciting benefit of cloud capabilities is the rapid/hyper agile development and deployment of services for a variety of services to meet evolving/dynamic needs. In particular, using loosely coupled services in a cloud based on established standards can not only expedite the availability of new services, but will also open the door for re-using services developed for other users (military, commercial re-use options).

With the cloud(s) in place, there are a number of emerging technologies that will be of interest to local governments as well as residents. In many ways, the neighborhood watch of old is being replaced with an emerging cyber-watch community for today. There are a number of service compiler approaches that will further enhance a community's ability to leverage/exploit proven services developed by other local governments. Compilers enhance the discovery and exploitation of services in use by other communities and/or residing in open service repositories, thus further expediting and reducing the cost of service improvements. Additionally, there are many artificial intelligence-based services that enhance an end user's experience in using a community service by providing interactions tailored for a specific user via reasoning engines. Personalization of services and service access for specialty users (such as visually or hearing-impaired residents) can continue to improve and extend services to all residents. And finally, there are numerous analytics that help local governments digest massive amounts of data. Data overload is perhaps the most limiting risk for any agile government project, and there are proven analytical processing services that run in a cloud to mitigate this problem. A layered analytics framework with loosely coupled plug-and-play analytic code is an emerging model of preference at all levels of government.

Today, many local governments are already overloaded. Our experience has been that any step toward agile government must begin with a recognition and appreciation of the heavy demands and limited availability of resources for government personnel who support emergencies and who manage acquisition. Likewise, communities must collaborate and leverage/re-use a much higher percentage of core-enabling technology developed and shared by others. Local governments must, likewise, leverage their technologies—including cloud-based solutions—with a focus on the agility of any local government to respond to the ever-changing dynamic needs of a community.

Each of the book's chapters focuses on some of the most contemporary issues facing public safety and local government. With so many contributing authors, from many different locations and perspectives, we have intentionally maintained their original voices to preserve their full intent and meaning. This book deals with issues regarding leadership, governance, broadband, interoperability, risk assessment and management, security, mobility, innovation, social/civic media, collaboration, procurement, architecture, as well as best practices, and more.

In contemporary society, there is no shortage of information and information technology—in fact, we are inundated every second of every day—from every conceivable source. Our challenge, then, is to make sense of what is hidden in the unprecedented amounts of digitized voice, video, and data. We must continue to seek innovative ways to better involve our citizens whenever possible. Finally, local governments must continue to seek out ways to further the training and development needs of its leaders in the appropriate use of technology systems and, as importantly, the human network, too.

Alan R. Shark
Executive Editor and Executive Director, Public Technology Institute

Foreword

BY ZACHARY TUMIN AND WILLIAM BRATTON

From social media to mobile platforms to the cloud, everyone and soon, every *thing*, will be connected in ways unimaginable 10 years ago. Big data is here, geospatial has arrived, biosensors are everywhere, and rings of lenses and steel guard our cities. Information and communication technology—gear, infrastructure, and applications—is today reshaping the public space, the marketplace, the political race—anywhere advantage is sought.

Technology does that by creating new capacities to see, engage, and act—now with massive scale and with pinpoint precision. Identity, for example, is now a matter of geolocation—whether by DNA, face recognition, or scent-sniffing nanobots. Crowd-sourcing moves battlefield innovation fast from company commanders just leaving theater to those arriving next week. We mass-produce unique solutions for anyone—whether death from above by drone, or life-saving medicine tailored to your genome. Common operating pictures stitch together fragments of views, whether of the city block or the patient-in-care so that safety or health is possible as never before. And it's all in real time—technology can today deliver global situational awareness, from big picture to sensor image straight to the bridges of combat ships, the hips of agents moving the president through urban canyons, or the firefighter responding to high rise catastrophe or mega-forest fire.

At the center of it all is the new, digital-ready, always-on citizen, friend, worker and adversary. Armed with devices, he's brimming with highly-empowered personal agency, capable of nearly instantaneous search, find, and discover. She's networked by smartphone or tablet to tens or millions of like-minded friends, colleagues, or partisans. Together they're reshaping organizations as employees and bending them as consumers, spurring rapid innovations that scale and dominate.

Therein lies the tale. For great as the power is of technology to give advantage, in itself technology is no silver bullet. If it were, the one with the best toys would always win. Increasingly commoditized, innovated, and brought to market, however, technology for a fee lets anyone create effects better, faster, and cheaper—whether the latest in improved explosive devices, or the interoperation of dozens of enterprise databases.

Creating effects is one thing. Delivering results another. *Getting* to interoperability is no mere technical challenge, for example, but a deeply political one filled with negotiation and persuasion, threats and inducements. Assuring outcomes *from* interoperability—returning value on the investment—is no mere technical challenge, but the hard slog of management, and leadership.

In this age of rapid technical change, entrenched legacy systems and bureaucracies, and high personal agency, success depends on the ability to get people, networks, organizations, and technologies all working together, fast. More than ever, that means reaching across boundaries of government, industry, and citizens to exchange assets all need but which only some have—whether those are assets of technology or information, skills or authority, or financing and political support.

No one has it all. No one can go it alone. It takes *collaboration*.

How does collaboration happen? We can point to eight critical factors.

- First, people must align on a vision—aspirations they share. A common vision, simple in the extreme, is often easiest to attain and hold.

- Second, the path forward must have right-sized, right-shaped problems that people can make progress against fast—not necessarily solving all on Day 1 but delivering value right from the start and showing everyone—friends and foes—"This can work."

- Third, platforms are essential—"clearings" with rules and infrastructure that can support the exchanges of data, information, imagery, or other assets. Whether it's an operations center or a computer server or jungle clearing, collaborators require access to trusted, secure platforms.

- Fourth, collaboration has to pay. Everybody asks, "What's in it for me?" Collaboration can pay in many different currencies—money, power, prestige, or access, to name a few. But it has to pay. And pay well—be prepared to prove that what you offer is exponentially more valuable to partners than what they now have.

- Fifth, the right people matter. The people you recruit send strong signals about the seriousness and capability of your collaboration. You'll need executive cover and technical prowess; you'll need people who can provide financial and political support. Others will be waiting to join in—if they think the collaboration is a winner.

- Sixth, performance. Be prepared to test and prove your results—and then communicate your achievement. Folks who stand to lose by your success often know who they are far better than folks who stand to win. Deliver—and then communicate.

- Seventh, stay in your political headlights. Don't race too far too fast, or hang back with over-caution. Don't stray left or right. Find the sweet spot of support and stay in it for the duration—assuring your support from start to finish so that you can deliver on the promise.

- Eighth, provide leadership that pushes the disruption forward—adjusting the vision to align for all as new partners come in, shaping and sizing problems just right, keeping the right people involved, holding political support, delivering results, assuring collaboration pays.

Each of these factors is a risk to be managed across the collaboration life cycle. Technology helps, and is essential: it can take the technical friction out of information sharing, for example, and make possible never-before-had situational awareness. Collaboration can get the necessary vision, plans, people, and support lined up to achieve that. But stitching it all together to deliver on the vision of a better day, the promise that garnered risk-taking, investment, and support for greater safety, or improved health, or more delighted customers, takes strategic management across the life cycle of collaboration. That is the great task of leaders in any age, but in our world today, where the race is on to out-compete, out-innovate, out-advantage—out-collaborate - surely more than ever.

ZACHARY TUMIN leads the project in Information and Communications Technology and Public Policy at Harvard Kennedy School. In addition, he directs the Harvard component of the joint Harvard-MIT initiative in cybersecurity.

WILLIAM BRATTON is the chairman of Kroll Advisory Solutions. He was police commissioner of Boston and New York City, and police chief of Los Angeles.

Tumin and Bratton are co-authors of Collaborate or Perish! Reaching Across Boundaries in a Networked World *(Random House, 2012).*

Acknowledgements

My sincere thanks go to the leadership and staff of the Public Safety Foundation of America as well as APCO International, whose support made this publication possible. PTI enjoys a long-standing relationship with both organizations. As with any edited work, we are indebted and grateful to the 36-plus contributing authors for their time and excellent contributions.

Thanks must also go to PTI's Public Safety Technology Council for their endorsement of the project and to PTI's Deputy Executive Director for Research and Government Services, Ronda Mosley. Ronda also serves as the staff liaison to the Council.

Dale Bowen, PTI's Deputy Executive Director for Program Development, was instrumental in the initial reviews of many chapters and played a leadership role in identifying and working with some of the authors.

I also want to thank an old friend of PTI, Alan Leidner, who currently works with Booz Allen Hamilton, for his continued leadership and advisement on matters pertaining to every aspect of GIS. In addition, I need to recognize Dalton Pont from SRI International, who has kept me in the loop with some incredibly cool and vital technologies that we will be seeing in the public safety space very soon.

We are also grateful to the Public Technology Institute's publishing team, which includes Lindsay Isaacs, who served as the project's copy editor, and Sally Hoffmaster the graphic designer, who designed the cover and laid out each page of the manuscript. They put in many long hours accompanied by many changes along the way.

Recognition and thanks must also go to former APCO International Executive Director George Rice for his vision and support, and who was our initial supporter of this project.

Chapter 1: Policing in the World of Social Media—A Guided Tour with @PhillyPolice

BY KARIMA ZEDAN, CHARLES H. RAMSEY, NOLA JOYCE

"History will be kind to me for I intend to write it."
    —Winston Churchill

We live in an increasingly complex and globally connected digital world. By 2017, 85% of the world could be connected to high-speed mobile Internet, according to a recent projection by Swedish phone company Ericsson (June 6, 2012). Facebook, the massive social networking service that has changed how individuals, organizations, and governments can interact with each other, has more than 800 million current active users and is soon to surpass the 1 billion user mark as early as August 2012 (iCrossing projection, 1/11/12). Right now, as you are reading this chapter (if it's 2012), YouTube users are uploading 60 hours of video to this Google-owned video-sharing platform every single minute. Social statistics like these can be overwhelming and confusing. It's an entirely different informational and technological environment for police departments than it was five years ago. Like much of government, police and law enforcement organizations tend to lag behind contemporary cultural trends, and we find ourselves searching desperately to understand how we can participate.

The good news for all police departments is that there is no expiration date for joining the social media conversation. The bad news, if you're communications-averse that is, is that you will have to join this conversation, because these conversations are happening with or without you. By the end of this chapter, however, we hope that you will be less skeptical and more curious about what this world of new media can offer you, your organization, and the communities you serve.

Our goal is to share our experience engaging in the world of social media at the Philadelphia Police Department. We are not speaking with a singular voice, imposing our norms and values for all of policing to follow. We are simply offering our perspective and the insights we've learned over the past three years as engaged participants. We are not claiming to be experts in this field, and if anyone presents herself as an expert, we would advise you to proceed with caution. There are no experts in this new world, only students and practitioners, and you can become both of them.

Don't ignore the obvious...

When thinking about social media and policing, it's important to pay attention to the actual words. "Social" says it all. This isn't a one-way, one-dimensional, top-down transaction between government and its citizenry. It's social, meaning human, relationships are at the core of the interaction. We become "socialized" when we learn the skills, knowledge, norms, and values that help us participate in our groups and our communities. Social media is dynamic and collaborative by definition.

Communicating in this world requires a social mindset, which in and of itself can be challenging for the often hierarchical, quasi-military culture that structures most police departments. Communication inside of a police department has traditionally followed a top-down or a bottom-up (through the chain of command) process of developing and pushing out highly controlled messages. It is often faceless, written, highly formal, and for the most part, informational and technical.

If this is your first foray into social media, you may need to un-learn and re-learn a few principles of communication. The first is that once you create the message, you cannot control how it is interpreted.

In the social world, the message will change in unpredictable ways the very millisecond that you press the return key on your computer or mobile device. Within minutes, your social message will be read by potentially thousands of people, with each person interpreting it according to her perspective and value system. She will then read the message and forward it, or "share" it, with her friends, family, peers, anyone, and everyone in her social network. This is both the beauty of social media and the sheer apprehension that it may inspire. The effects are exponential and difficult to measure. You must be willing to let go of the notion that you can control its interpretation; you simply cannot. But then again, you never could, and if you thought you could, it's time to take a second look at how messages actually get communicated.

Take the Philadelphia Police Department. We are the fourth largest police department in the nation, with more than 7,200 sworn and civilian members. Messages that start from the very top, at the level of Police Commissioner, must travel through seven layers of rank, across 140 square miles, in 55 different facilities, before reaching the police officer in a police district. Communication looks more like the childhood games of "whisper down the lane" or "telephone" than it does the one-way exchange between sender and receiver in which what is said is actually aligned with what is heard.

During transmission, the message will morph into a different form and be changed by many pressures, people, other groups, including unions, traditional media, and perhaps the most powerful influences of all, rumors and stereotypes. By the time the message actually reaches the police officer, she will hear it, process it according to everything she believes, ask her peers, run it by her supervisors, and ask questions such as, "What does that really mean?" The bottom line: What the receiver hears will most likely be different than what is intended by the sender. This is already happening right now in your police organization. Communications are being interpreted in so many different ways beyond your control.

One of the primary differences between traditional methods of communication and social media, however, is that we are engaging in a conversation, and over time, we are expressing our individual and collective personalities. Monologue has given way to dialogue, and in this new social landscape, there are endless opportunities in which we can create positive contact with not only the public, but with our own people, as well. Social media levels the playing field for government and for our communities. For those who are willing to adapt to this environment, the playing field is rich with opportunity.

If it bleeds, it leads....

At this point, we need a solemn reminder of the status quo. All of the below headlines were taken from the local media in Philadelphia, namely our print and digital papers, the Philadelphia Inquirer and the Philadelphia Daily News. This is what the world looks like when police departments sit on the social media sidelines:

- "Three people shot in night of deadly violence, police investigate multiple shootings."

- "Homicides in Philadelphia continue to rise."

- "Accused officer's home searched."

- "Philly cops beat 18-year-old for running stop sign."

- "Girl, 3, grazed by stray bullet in West Oak Lane."

"If an idiot were to tell you the same story every day for a year,
you would end up believing it."
— Horace Mann, American education reformer

If you were to glance at the headlines stemming from the traditional media in today's news, would they look somewhat like the headlines written in the above section? If you have a moment today, and if you haven't already, turn on one of your local news stations at 5 or 6:00 this evening, or glance at your local paper online, or to get a sense of what is happening nationally, go to usatoday.com, or turn on CNN (which also live streams its content on cnn.com). What will be the content generated from your local news affiliates and organizations?

As journalism aphorisms go, "If it bleeds, it leads" is still at the top of the list. What constitutes the selection of content in the news is a choice. It is a choice by media organizations as to which stories they will report on a daily basis. It is a choice driven by a number of factors that are not the subject of this chapter. Their choices exert great power and influence over your organization's reputation, the image of your city or town, your sense of credibility, organizational effectiveness, and sense of mission and morale.

As a police department, we have an absolute responsibility to provide the news media with as many answers to their questions as we can. This is a necessary part of our mission, to give accurate and timely information about all things crime- and public safety-related. Transparency with and accountability to the media are vital to the health and well-being of our organization and the governments of which we are a part. We must always strive to uphold the integrity of our organization by responding to media requests, and providing as much open access to data, statistics, and our own people as possible.

If we continue to rely only on the media to communicate police-related stories, however, we will be waiting a very long time before they are interested in covering anything that doesn't involve guns, drugs, and corruption. And by very long time, we mean it will simply not happen.

Tell your own stories

Enter the world of content marketing. Content marketing is a term derived from the private sector that encourages companies to create content that reflects their brand, their service, and/or their product, and share it directly with their customers. There's nothing mysterious or glamorous about this term. It's a way of thinking, a driving philosophy that empowers us, as police departments, to tell our stories, on our terms. Ten years ago, telling our stories and finding a platform in which they could be distributed was challenging.

Times have changed dramatically, though, and so should we. Social media provides the tools and the methodologies to make content marketing a living, breathing, 24/7 reality for all police organizations. In this reality, there is a robust market for self-promotion. In other words, we are not constrained by the traditional media to report on positive, interesting, and engaging stories about the people who actually make our police departments work. We don't have to "pitch" a story into the void of traditional news media outlets and hope that a slow news day will work to our advantage.

In this new media reality, we can create these stories and share them with the world. No outside media is required to make this happen. In doing so, we are not only reaching the communities we serve in a different way, but we are demonstrating to our own people that we, as an organization, value their efforts and their contributions to the mission. This is not a small act of recognition. When we post a profile of one of our "Philly's Finest" on our website and on our Facebook page, 40,000 people in Philadelphia and around the world may view it during the course of one week. We are putting a human face on the large, complex, and bureaucratic machinery that is the Philadelphia Police Department, and that is the stereotypical image of city government. The more we personalize our organization, the more that we can become known and integrated into our communities. Telling our stories socially is another dimension of good old-fashioned community policing.

Fighting crime socially

If positive, human-interest stories don't move you, what about using social media platforms to enlist the public's help in fighting and solving crime? In reality, of course, it's not an either/or proposition when it comes to how police departments can use social media to engage their communities. The tools are there for the taking; use them freely and creatively to connect to the world.

In our social reality, sharing photos and video rank very high after entertainment and gaming mobile apps. Snap a picture, and within moments, you can upload it to your favorite social networking site to share with your favorite 500 friends. In the recent past, police departments have depended on traditional media for distributing pictures of wanted suspects, recently arrested suspects, and video of criminals caught on tape.

Web 2.0 technologies and social media platforms like YouTube and Facebook, however, have transformed how police departments can get this type of information to the media and the public in real-time. We no longer have to rely on the digital infrastructure of television and radio media outlets to broadcast photos and videos. YouTube, Facebook, and Twitter (and a host of other sites) are free platforms. If your organization doesn't have a modern website, you can use social media channels to post suspect photos and surveillance video. Practically speaking, police departments can distribute photos and video anytime, from any device, on a 24x7 basis. If there is a dangerous suspect at-large, we can ask the public for help immediately upon receiving that information.

Case in point, on February 10, 2012, Marcise Turner was arrested for the sexual assault of a 9-year-old girl in Philadelphia. Special Victims Unit investigators were dispatched to the scene at 3:45 p.m., where they interviewed witnesses and viewed surveillance video that potentially captured the suspect on video. At 5:15 p.m., a photo of the suspect is retrieved, and at 5:50 p.m., that photo is then forwarded to our Office of Media Relations and Public Affairs for distribution on our website, phillypolice.com and YouTube channel. At 6:20 p.m., a concerned Philadelphian calls the Police Department with information about the suspect, and within five minutes, police officers arrest Turner at his residence.

This example is just one of many that demonstrates the power of social media technology in service of our crime-fighting mission. The Philadelphia Police Department has distributed countless photos of wanted criminals and over 250 surveillance videos with suspects who have been recorded on camera. For the latter alone, more than 30% of the surveillance videos have resulted in an arrest of a suspect. That is a conservative estimate, as we cannot always account for the source of a crime tip that comes into the department. The more people who can access crime information directly from their smartphone or their computer, the more likely that we will be able to receive tips that could lead to arrests.

Traditional media no longer has to serve as the intermediary between the police and the communities we serve, though of course, they will remain an integral and very helpful part of the process. Again, it's not an either/or proposition. Police departments can harness the power of social media and mobile technologies, and use them as a force multiplier to reach community and police personnel alike. These tools add another dimension to how and why we communicate with the world. They are not a substitute for the necessary face time with traditional media outlets via press conferences and interviews. They are supportive of everything that you are already doing as an organization in terms of your media relations and public information outreach.

Keep the big picture in mind

The importance of having a technological infrastructure that supports our social media efforts cannot be overstated. When the department launched www.phillypolice.com on December 30, 2009, we officially stepped into the digital age with the right gear to navigate this new terrain. Having a website with a convenient and easy-to-remember name allows us to store all of our in-house generated stories and public safety information in a single location.

We use phillypolice.com as a home base for everything department-related, and we link to it from our social media channels. Irrespective of what social platform we use (Facebook, Twitter, YouTube or Google+), people can always return to our website as the official source. As the social world evolves, the number of different types of platforms will undoubtedly increase. It is incumbent upon us, as government organizations, to stay abreast of current trends, popular platforms, and emerging technologies. Right now, Facebook is very hot in the world of social media, but five years from now, there will most likely be another platform to which we have to adapt.

We have found that there are a few larger themes that cut across all platforms and media. Here is some modest advice and lessons gleaned from our own experience as a large urban police department jumping into the social mix over the past three years.

- **Practical advice:** Though this chapter has not been focused on the "how to" of social media, or the detailed tactics of developing a communications function in your organization, we would be remiss if we didn't state the importance of establishing reasonable policies and procedures. This is new terrain, and you will need a solid framework in which to move forward with your efforts, which includes social media policies, media relations policies, proper internal organizational procedures and protocols, and training for personnel who will be implementing your efforts. There are a number of resources that can guide your organization in developing the right infrastructure for policies, including the International Association of Chiefs of Police (IACP), networking with other police departments, attending professional conferences, and consulting with your local law department, which most likely has already or is developing a social media policy for your municipality.

- **Stay open-minded and flexible:** This may seem like very banal advice, but it will be extraordinarily important as you move through the highs and lows of exploring the world of social media. Hire curious people to guide you through your social media efforts, and stay curious as a leadership team. This is uncharted territory, so mistakes will be made, and this is not a panacea for all of your communication challenges as an organization. It is a space that is filled with possibilities, and the rapidly changing nature of technology offers you something new at every turn. You have to challenge your organization to move beyond the status quo, away from your comfort zone, and be open to exploring what this world can offer you. It's easy for police departments to be apprehensive of any new communications/media platform that could potentially be misused (that's part of our culture) or which will draw out the most cynical of people whose sole purpose is to say something negative and generally offensive. This will happen with or without your participation, and it's already happening now. Quite simply put, you can't control the haters, (remember, you have to let go of that illusory sense of control). It's tempting to lead from a place of fear, but if you do, you'll be limiting what you can achieve. Try to stay as open as you can to discovering something new, something helpful, something that will create a positive experience for communities and police alike. What you can control is the effort that you put into crafting your message, telling your story, and putting reasonable policies and procedures in place that will allow you to respond appropriately when people make mistakes.

- **Put people in charge of your efforts who demonstrate extreme curiosity:** If the people behind your social media communications efforts are mandated to do this job, it won't give you the best possible outcome. Be thoughtful as a leadership team, and choose people who really want to do this kind of work and who are predisposed to learning (cynics, curmudgeons, and self-promoters need not apply). These people will be your representatives for your organization in this arena. Are they trustworthy? Do they demonstrate good judgment? (This is key, as it's a new area.) Can they problem-solve on the spot because social media engagement happens in real-time? Do they understand the mission of your organization and your personal vision as a leader? Curious, inquisitive, risk-takers (within reason) who are committed to the mission are good candidates to help you turn your social media efforts into a meaningful endeavor.

- **Internal/external is becoming a false dichotomy:** The lines between public and private and internal/external are being completely blurred. Everyone will have access to your social communication, which includes media, community organizations, other city government agencies, and your own people. Don't ignore the many opportunities to use social media channels to connect directly with your own employees. They will benefit from reading positive stories in which they can see themselves as goodwill ambassadors for your department. As an example, our internal communication platforms are not well developed. We can bypass our impoverished internal infrastructure, however, by posting everything from surveillance videos to unit profiles on phillypolice.com. Most of our personnel can visit the website directly from their smartphones, which in this day and age, most of our police officers own, or any computer with Internet access. Promoting our external website has become a valuable communications tool for reaching our own people.

- **Storytelling is mission-critical:** Telling stories is one of the most effective methods of communication because they are easily remembered and motivate people to think and act differently. Rather than detail a set of facts, figures, and orders, as is the nature of most internal police communication, use stories to convey critical messages that will really stick with your audience. Just ask yourself, if you are listening to a public speaker, what are you most likely to remember: a series of statistics from a PowerPoint slide or theoretical projections about future outcomes, or a story that conveys the knowledge learned from an interesting past event and inspires you to act differently in the future? Pay close attention to every time you now have to listen to someone speak, and see how often he/she uses stories to convey the message.

- **Lean forward but don't lose your balance:** Success in any arena doesn't happen overnight, and social media engagement is no exception. When we as a department established our Twitter account, we didn't engage in any dialogue for three months. We watched other people, listened to the conversations that non-police organizations were having, and studied the "Twitter-verse." We learned about what people thought was interesting, exciting, and funny, and what moved people to respond. We were patient. Our first interactive tweet with someone involved responding to a comment about an abandoned car that had been in her neighborhood for months. We sent her a message giving her the telephone number to our Neighborhood Services Unit and told her to take down the make, model, and license plate. 15 seconds. That's all it took to respond to her, at no cost, and in a customer-service friendly manner. She was grateful, said thanks, and moved on to another issue. Though it might be tempting to start engaging right away as soon as your organization sets up accounts across the many platforms available, it will only benefit you to be measured, and get your finger on the pulse of what your communities want from you. Tread lightly and deliberately, and over time, your comfort level will grow. Learn as much as you can from how other organizations are using social media to start a dialogue. Look around at powerful businesses with strong social brands, and read their conversations on Facebook and Twitter. Examine how other government organizations across the country are adopting social media into their community engagement and civic participation strategies. Study how and when they do it. This is an interdisciplinary endeavor. Take advantage of the wealth of material that already exists and examine how you, too, could apply it in your strategy.

- **Develop your brand:** Organizations are being turned into social personalities through their digital presence—they are creating their own "brand." Brands are not just for corporations like Nike, Coke, and Target. Every organization, irrespective of the industry or sector, carries with it a number of different symbolic attributes. Your brand isn't what is represented on your logo, conveyed in your mission statement, or in the image you are trying to project, though all of those things definitely contribute to your brand. It's more than a singular identity. It's a dynamic, living, and breathing thing that evolves over time that can both influence and be influenced. When you think of the FBI or the Navy Seals, what comes to mind? What about when you think of BMW or Apple? Each organization evokes different feelings or thoughts about its quality, its integrity, and its value. When enough individuals begin to think the same thing about an organization, its brand can shift in any direction. As police departments, we can tell ourselves time and time again that we are trustworthy, credible, professional, even outstanding, and best in class, but what does the public think? Would they agree with us?

Your brand is not just what you say you are, it's also what they say you are. We are one of the most public-facing organizations in government, and police departments are often the first, sometimes, only contact that the public has with government. What will that person say about her police department when the interaction is over? Whatever she says, it should matter. It's not all that matters, but it matters. The world of social media can be used to create a different kind of touchpoint for police departments and the communities we serve. Over the course of hundreds and thousands of interactions, we will begin to shape our image and reputation. It could be for the betterment of your organization, but you will have to, as Twitter implores us, join the conversation to get the ball moving.

- **The Positivity Ratio:** Positive community contact only takes a few seconds in the social media environment. It takes many more positive interactions with police officers/organizations to counter the effects of a single negative interaction (we've heard various numbers, such as 13 positive interactions for every negative interaction, or 10:1 ratios, or 5:1 ratios). Whatever the ratio actually is, we know that we need more positives to offset the impact of one negative (such is true in all facets of life, as well). Negative experiences impact the part of our brain which is wired for remembering and easily conjuring up the powerful emotions associated with it. The world of social media offers us an avenue to create more positive experiences for our organization. Why not lead the way and create positive experiences where none has existed? Opportunity is knocking; are you available to respond?

A few final thoughts…

There are many contemporary issues in policing that demand our immediate and future attention if we are to stay relevant and capable in our mission. More often than not, we struggle to keep pace with the ever-changing social, political, and technological environment. Policing in the 21st century requires a mindset that is wired for managing complexity. We hope that we've conveyed that the world of social media is filled with possibilities for police departments even amidst this complexity.

Experience has shown us that not every tweet, Facebook post, or video will be received in the way in which we intended it; that is to be expected. Every time we step forward, however, we learn something about what sticks, and what moves people to want to make contact with us. It's a large-scale experiment, which will make all police departments slightly uncomfortable. Policy, procedure, and training matter greatly; we're not undercutting their importance. They provide the structure necessary in which to have the flexibility to respond. You need both in order to thrive within the complexity. Become a seeker in this social land. Explore, participate, advocate, educate, organize, and engage with a healthy dose of measured curiosity. You will undoubtedly be surprised at what you can learn and achieve.

KARIMA ZEDAN has served as the Director of Communications & New Media for the Philadelphia Police Department (PPD) under Commissioner Charles H. Ramsey since June 2008. She has worked with the executive staff to deliver a high-performing communications and media relations function and strategy. Prior to her tenure with the PPD, she worked in communications and external affairs at the Division of Public Safety and the Provost's Office at the University of Pennsylvania. She holds a Bachelor's in Communications from the University of Pennsylvania, a Master's in Criminal Justice from Temple University, and is currently a Ph.D. candidate in Criminal Justice with a focus on organizational communications and innovation.

Philadelphia Police Commissioner CHARLES H. RAMSEY leads the fourth largest police department in the nation with over 6,500 sworn members and 830 civilian members. He brings over 40 years of knowledge, experience, and service in advancing the law enforcement profession in three different major city police departments, beginning with Chicago, then Washington, D.C., and now Philadelphia. He is an internationally recognized practitioner and educator in his field, and currently serves as President of both the Police Executive Research Forum and Major Cities Chiefs Association. Commissioner Ramsey holds both a Bachelor's and Master's degree in criminal justice from Lewis University in Romeoville, Illinois.

NOLA JOYCE joined the Philadelphia Police Department in February 2008 and serves as the Deputy Commissioner and Chief Administrative Officer for the Office of Strategic Initiatives and Innovation. Prior to her arrival in Philadelphia, she worked at the Metropolitan Police Department, Washington, D.C. (MPDC) from 1998 to 2007, and guided the expansion of MPDC's community policing model, the alignment of the budget with strategic initiatives, and the implementation of significant changes in the department's organizational structure. Before her tenure in Washington, D.C., Joyce spent six years as the Deputy Director of the Research and Development Division for the Chicago Police Department where she was essential in developing and implementing the Chicago Alternative Policing Strategy (CAPS). Joyce holds three master's degrees and was recently accepted into Temple University's Doctor of Philosophy program in Criminal Justice.

Chapter 2: Public Safety Enters the Collective Intelligence Era

How high-speed wireless communications are helping first responders overcome Public Safety Enemy #1: The unknown

BY PAUL STEINBERG

- *Unknown to the officer, the driver she has pulled over for speeding has just been involved in a domestic violence incident involving a firearm.*

- *Officers responding to a "shots fired" call are unknowingly approaching a gang shootout on a route that could put them directly in the line of fire.*

- *As the first firefighters enter a burning warehouse, the fire is approaching a storeroom they don't know is filled with highly combustible cleaning supplies.*

It's often said that what you don't know can't hurt you. But in public safety, nothing could be further from the truth. In fact, perhaps the biggest challenge first responders face is dealing with the unknown. That means the biggest challenge a CIO or IT director faces is how to make sure the right information gets to the right people at the right time—how to ensure every police officer, firefighter, EMT, and dispatcher knows virtually everything important about every incident or situation—in other words, how to make the unknown known.

Information and communication

The answer, of course, is information—more specifically, the communication of that information to first responders to help them be more productive, more protected, and more effective. This is known as "situational awareness," which will be discussed later in the chapter. Over the years, both the level and the quality of available information and communication have changed substantially. In the 1990s, the push was for communications technology that could get information in people's hands to help improve personal productivity. This technology included personal computers and digital assistants. In the early 2000s, things changed. The objective became knowledge distribution, pushing information to everyone who needed it via connectivity to whatever mobile devices they had in their hands. By and large, this was a one-way data exchange from the command center to personnel in the field.

Collective intelligence

In the 2010s, we've entered the era of broadband-enabled collective intelligence. The idea of collective intelligence is that information is not simply pushed out to individuals in the

TECHNOLOGY TRENDS

- More than 35 percent of the global workforce will be mobile workers by 2013.

- People spend more than 700 billion minutes per month on Facebook.

- More than 30 billion pieces of content (web links, news stories, blog posts, notes, photo albums, etc.) are shared each month on Facebook.

- Every minute, more than 60 hours of content is uploaded to YouTube. In fact, more video is uploaded to YouTube in a month than three major U.S. TV networks produced in the last 60 years.

- YouTube is localized in 39 countries across 54 languages.

- Over 300,000 users helped translate the Facebook site through the translations application.

- Smartphones surpassed personal computers in units shipped for the first time in 2010.

- It took tablets less than two years to disrupt the commercial PC market.

Increasing complexity of consumer electronics means U.S. consumers are returning products even when they're not defective. Accenture found that 68 percent of returns are categorized as "no trouble found," meaning that the customer believed there was a defect, but testing failed to detect a problem.

---

field (like directed spokes from a hub), but shared among all team members and with disparate teams using two-way broadband communications technology. The goal is increased situational awareness. The reality is, each individual or team at an incident scene has a unique perspective on the location and can collect information that is unknown to others. Next generation public safety networks will make it possible for the "collective"—i.e., everyone on the team from dispatch to command to individual first responders—to both access and contribute to this critical information in real time.

Mission-critical social networking

Not coincidentally, the collective intelligence era correlates with the era of cloud computing and social networking. Both consumer and business customers are communicating and collaborating on sites such as Facebook, Twitter, YouTube, LinkedIn, and scores of others. How widespread is the social networking trend? In March 2012, Facebook had more than 845 million active users, over half of whom access the site on their smartphones or other mobile devices. YouTube is another example, with more than 4 billion video streams being accessed every day.

With collective intelligence, we are essentially taking the social networking phenomenon and applying it to public safety. The entire agency is communicating and collaborating in a social, interactive way over land mobile radio and broadband networks.

Using the "cloud" as a data aggregation point, the most relevant, up-to-the-minute information can be made available to everyone on the network in a secure environment. Public safety professionals can interactively collaborate over a variety of communications technologies to bring up-to-the-minute knowledge to the response team as events evolve and are resolved. The cloud is also beginning to play an increasingly important role in other applications, such as disaster recovery.

> Recent statistics from the Washington, D.C.-based Police Executive Research Forum tells the story:
>
> - 83 percent of police agencies use social media to share information with the public.
>
> - 70 percent use it to receive tips from the public.
>
> - 89 percent use it to monitor investigative leads.

Situational awareness

The idea of collecting and sharing information and intelligence is hardly new. As Bill Schrier, former CTO for Seattle, Washington, and a former police officer, notes, "Cops have done this forever, talking to people on the beat and to other cops, gathering and sharing information from a wide variety of sources. But now, we can collect and share this intelligence using new social networking tools that allow collaboration among both Land Mobile Radios (LMRs) and broadband devices, like smartphones and tablets." What's new is the ability to collect mission-critical information from a wide range of media and residents—from 9-1-1 calls to street video surveillance cameras to interactive maps to information from a myriad of databases to on-scene, on-person video cameras—then share it immediately with anyone and everyone, including residents en masse and in near real-time. It's all achieved with powerful, secure, high-speed communications networks and the social networking cloud.

"We want to be able to collect a complete picture of an event by amassing information from a wide range of sources," says Jeff Johnson, CEO of the Western Fire Chiefs Association. "We then share this collective intelligence by pushing the information back to the field." It's all about eliminating, or at least considerably reducing, the unknowns. The more public safety officers know about a situation—profiles and photos of suspects, streaming video of the actual location, area maps, building plans and more—the safer and more effective they will be, and the safer the community and its residents will be.

Collective intelligence at work

What does collective intelligence look like in the real world? A SWAT team dealing with a hostage situation, for example, can use two-way streaming video for communications and collaboration. High-performance, bi-directional communications allow the "feet on the street" to not just access the collective intelligence, but dynamically contribute content to it. Video is captured on the scene by officers on the ground, in vehicles, or helicopters. Cameras can be placed on mobile robots that actually go into the building and stream video back to command. These constant feeds of images and information from different perspectives are streamed to the on-scene mobile command unit. The team leader then edits the video feeds to provide responders with the most relevant information, and disseminates the video to whoever needs it—including sniper teams—to enhance safety by enhancing situational awareness.

Collective intelligence is just as critical on a fire scene, for example, at a chemical plant. To begin with, the incident command center and individual firefighters can access a wealth of real-time information on their wireless broadband devices, such as ruggedized hand-held computers. Each team member has immediate access to building details, such as floor plans, sensor inputs, location of electrical panels, and the types and locations of hazardous materials. They also can get clear audio communications, as well as streaming video from various angles and perspectives. When the incident commander in the mobile command center views live footage from a helicopter, he sees what the team on the ground can't see: that the cut they're about to make to vent the roof is being made in a less-than-optimum place. The commander streams the video to the team leader, who moves the venting operation to a more effective location.

Eliminating the unknowns

How does the concept of collective intelligence help the first responders in the examples at the beginning of this chapter?

- As the car is being pulled over, the dashboard camera scans the license plate, and the system automatically sends the data to the statewide database, revealing a long history of traffic offenses. Then, before exiting her vehicle, the officer receives information that 20 minutes ago the driver was involved in a domestic violence incident and should be presumed armed and dangerous. She calls for backup, and her in-vehicle workstation shows her when and where the backup will arrive. In moments, the team of officers has the suspect in custody. All the activities have been recorded and sent in real time to the command center and dispatch.

- Officers responding to the "shots fired" call can immediately access streaming video of the scene from street video surveillance cameras, including those of private businesses, gaining real-time situational awareness that enables them to take a safer route to the incident, avoiding any crossfire. A contingent of backup units also can access the video on the network. In addition, officers can access video of the fleeing suspects taken from a resident's cell phone, making it easier to identify suspects and make the arrests as they try to escape. No one is injured. Overlaying the first responders (vehicles and individuals) on a map or floorplan that can be seen by all participants allows for a coordinated and orchestrated plan to cope with the incident.

- Before entering the burning warehouse, firefighters can confer with responding utility crews regarding the building's electrical systems. They access current building plans both on their in-vehicle workstations and hand-held computers. They are made aware of the hazardous material storage location, enabling them to better assess the situation and quickly move to extinguish the flames before they reach the storeroom. With no explosion to increase the fire's intensity, scope, and peril, firefighters can contain the fire to a single floor with no serious injuries and minimal structural damage.

Collective intelligence and LTE

What makes these powerful collective intelligence capabilities possible? One key element is a dedicated high-speed broadband network—such as LTE—that delivers innovative and secure data services that complement and integrate with mission-critical voice over LMR networks. How will we achieve this? Today, there is strong momentum behind Long Term Evolution (LTE) 4G networks as the preferred broadband network technology for public safety use cases worldwide.

The emergence of LTE networks for public safety

"They can do that on TV. Why can't you do it?" Crime shows and police dramas are among the most popular on television. Many of them show officers using technology that is somewhat enhanced and futuristic for dramatic effect. Of course, when the public sees something on the TV show "CSI," they think just about any police force can do that, too. While this isn't yet the case, it won't be all that long before it is. And LTE is one of the enabling technologies that is getting us there.

LTE is a transformational standards-based technology for many industries. What makes LTE so important for public safety? Its powerful mobile capabilities extend ubiquitous broadband coverage, higher capacity, enhanced security, and high-speed two-way communications wherever they're needed in the public safety network. LTE's exceptionally low latency is key to managing and delivering high-performance solutions, such as video, telemetry, analytics, and a host of other advanced applications. Examples might include helicopter video footage of a fire scene, motion detectors that activate surveillance cameras in a high-crime area and real-time images from a remotely controlled robot sent into a possible bomb situation.

Collective intelligence capabilities

"Public Safety LTE is the next generation of wireless broadband technology," says Rick Keith, senior director, Private Broadband at Motorola. "It's designed to, at last, give first responders the ability to access mission-critical content whenever and wherever they need it." Just as they now have mission-critical voice on LMR, with LTE's broadband capabilities, they'll also have mission-critical data. These capabilities will be essential in daily public safety activities, delivering exceptional real-time situational awareness and allowing network managers to prioritize and pre-empt traffic whenever necessary in emergency situations. In addition, LTE is optimized for multi-jurisdictional emergencies such as intra-state criminal activity, cross-jurisdictional high-speed pursuit, and widespread natural disasters, ranging from tornadoes to tsunamis.

"Public Safety LTE helps create an interoperable system that many different agencies can work from," continues Keith. "Multiple agencies are able to jointly respond to a variety of incidents both major and minor." It's quickly becoming clear that LTE is the technology that will allow first responders to take maximum advantage of innovative devices and creative applications that will empower them to be more focused, more efficient, and more effective in their mission-critical public safety activities.

Coverage, capacity, and control

Coverage, capacity, and control are also major network considerations. To begin with, no one can predict where the next crisis incident will occur. For most agencies, this means different coverage requirements than public carrier systems. When the unthinkable happens—a hurricane, a bridge collapse, a terrorist attack—public carrier networks may be damaged or become immediately saturated with civilian calls. Capacity is maxed out almost at once, severely restricting communications of public safety agencies forced to rely on them. Private networks reserved for public safety use give agencies the ability to control who gets on, improving the response during disasters. Agencies also can give top priority to those responders who need information immediately, a capability not available on public networks. Public safety bandwidth needs differ from those of engineered commercial networks in that incident scenes create very localized high bandwidth consumption (often localized over an area of only a city block or less), and one cannot predict where these situations may arise.

Dynamic prioritization

Why is prioritization of one call over another so important? Let's say two data session requests come into the network at the same time. One is associated with an urgent "officer down" incident; the other is for dealing with a cat stuck in a tree. A public carrier network doesn't know that the "officer down" data session must have priority over the "cat session." A public safety computer-aided dispatch (CAD) system on a public safety LTE network does. It can automatically instruct the network to give the "officer down" data session priority, saving seconds that may help save a life. If you're using a public carrier network operating with commercial network priority policies, you don't have the ability to prioritize incidents. And the fact is, if a commercial network were responsible for prioritizing first responders, it also implies the possibility of denying or even preempting service to residents who may have a legitimate need to call for help or provide vital information to public safety officers.

Rishi Bhaskar, vice president of Motorola Solutions, explains public safety LTE prioritization succinctly. "Prioritization allows us to channel bandwidth or create larger bandwidth pipes specifically for people responding to an event at any given time. The key is dynamic prioritization," Bhaskar continues. "There will be times when we will need to increase a firefighter's throughput as he or she is responding to an incident, ensuring we are able to continually monitor their biometric data. Simultaneously, we may need to pre-empt another user off of the system to ensure the appropriate bandwidth is available in real time. It's all about prioritization that leverages public safety software applications for standard LTE networks and devices."

Public-private alliance

For today's new collective intelligence communications model, it's usually preferable that the public safety LTE network be a privately owned network rather than a commercial carrier network. Although carrier networks certainly can play a role in public safety, private networks allow public safety to define and build a network with the coverage, capacity, and control that first responders need in the heat of the moment. This ability to prioritize and control communications is only likely to be available with networks built for public safety. There's also the need for more hardened, i.e., more secure, networks to ensure continuous operation under adverse conditions and in hostile environments.

It's key to have a dedicated, prioritized, and preemptive public safety network you can count on to deliver mission-critical data in crisis situations, whether on a large or small scale. But public carriers also can enable crucial communications when our users are out of range of the private network or have the potential to provide additional capacity to augment the dedicated network. This is the public/private partnership the industry has been talking about for years.

Public carriers also have physical assets, such as cell sites, towers, and backhaul that can be cost shared with the public safety network. This is a win/win situation for both public safety and the carrier in that it shares the costs of these very expensive CAPEX and OPEX components of the network. This is one reason Motorola has partnered with Verizon to create nationwide interoperability, allowing public safety users to communicate even when they are unable to access the private network.

Collective intelligence and standards

LTE is a standards-based technology rooted in the 3GPP standards body. This standard is vital to the collective intelligence concept in public safety because it provides the ability to interoperate across multiple networks—a critical element for ensuring effective emergency response.

Standards also ensure robust, field-tested performance and help provide economies of scale that result in substantial equipment cost benefits. In addition, standards ensure that LTE will not be a one-time approach, but will be a never-ending roadmap of new features and applications developers can overlay on top of the standard, giving the network exceptional sustainability and scalability.

Because broadband can deliver so many different functionalities, none of us can anticipate exactly what innovations will be forthcoming. Some we already know; for example, applications such as eCitation, automated license plate recognition, and field-based fingerprinting for positive ID. But many other innovations are on the drawing board or exist only in the minds of the developers. One thing we do know for sure is these innovations will be happening sooner rather than later, and many of them require a high-bandwidth wireless network. We also know standards-based LTE will be able to support these new applications as they are developed. Another thing we know for sure, at least according to Gartner in a 2010 report titled "Debunking the Myth of the Single-Vendor Network" and other analysts, is that there is a trend away from single-vendor networks to multi-vendor networks. This is a trend that helps both enhance safety and reduce cost, and it's made possible largely by adherence to standards-based systems.

That's why it's important to always maintain the integrity of the standard, not only in the infrastructure, but up through the core and in all devices the network must support. This will ensure that your agency or department will be able to quickly adopt the innovative new devices and applications that enable police, fire, EMS, and other first responders to have life-saving intelligence and situational awareness in every incident and on every call.

Intelligent mission-critical devices

Intelligent mobile devices are crucial for public safety collective intelligence. LTE is a technology that allows first responders to take advantage of innovative devices and creative applications in a dynamic mission-critical environment, allowing them to focus solely on their mission. This means public safety agencies wanting to invest in next generation systems have to consider much more than infrastructure—this could include dongles, vehicular modems, and ruggedized smartphones or portable devices. They must also look closely at the profile of devices and applications they plan on utilizing. And they must consider them not just for today but as they will evolve over the life of the network.

Reliability rules

On the device side, the first considerations are ruggedness, reliability, and suitability for mission-critical work. Although some departments try to make use of consumer-grade mobile devices, such as smartphones and tablets, most end up being disappointed in the devices' ability to perform in the real world of public safety. Consumer mobile devices aren't built for being dropped during a foot pursuit, for being easily readable in blazing sunlight or operated with gloved hands, for enduring the heat and smoke of a smoldering stairwell or for withstanding the stress of a hazardous material cleanup in a downpour or blizzard. In the often-challenging and ever-changing circumstances of public safety professionals, ruggedness and reliability are crucial, both in connectivity and in the physical strength of the mobile devices being used.

The overall market trends toward increased use of smartphones, tablets, and other mobile devices are well documented, and devices like these certainly have a place in public safety.

But in most cases, we need to toughen them up considerably, and make them easy to use even in the most stressful and dangerous situations. But, there is still a lot that we can and should leverage from commercial devices, including their digital "heart" (CPU, memory, and peripherals), operating system, and even some applications.

Device innovation

In terms of the devices themselves, we are now only scratching the surface of what's possible. There are powerful in-vehicle computers, which have been in use for some time now. There also will be a new breed of ruggedized LTE handheld devices that will allow two-way communications that help gather and disseminate information to populate collective intelligence. They will keep first responders connected to dispatchers, the command center, other agencies, and everyone else needed to help resolve situations as simple as a traffic stop or as complex as an active shooter situation or a five-alarm fire.

Next generation public safety-optimized handheld computers will give first responders many innovative new capabilities. First of all, they will be ruggedized, with shock-resistant glass and hardened cases, so that when you drop them, they won't break. They will have touch screens that are a combination of resistive and capacitive, so they work in the elements. They will be wide-area enabled and use the 4G broadband network to provide real-time commercial grade push-to-talk and VoIP calling; they also will offer integrated contacts and easily accessible incident overviews. They will allow field officers to see assignments and relevant details of incidents with just the push of a button. They will provide email support and, if lost or stolen, offer ultimate security with remote lock and wiping of the SIM card, contacts, and content. In addition, they'll also feature intuitive, ergonomic design that makes them second nature to use in critical situations. Most of all, we want to ensure that the devices present a standard and open software platform that allows applications to be created freely, thus encouraging innovation to come from all quarters: industry, government, entrepreneurs, and universities.

Experience-based design

The focus today is on what we call user experience-based design, which includes the ergonomics of the platform, physical functionality of the device, and the user interface—in short, the way the user interacts with the device. Our Motorola Solutions design team has gone into the field to do primary research with first responders. As Rick Keith expresses it, "It's less about the technology and more about what our customers really need." The objective is to make each device as useful as it can possibly be, organizing functionalities in a way that the user experience leads the responder logically through an incident.

The top level screen, for example, can provide all the available assets and information associated with the incident. The responder can simply click to get data on suspects, like photos and rap sheets. They can click to pull up maps or building plans of a location as well as the locations of units already deployed. They can pull up available tactical video of the scene from either street or vehicle cameras in the area. They also can literally create a local

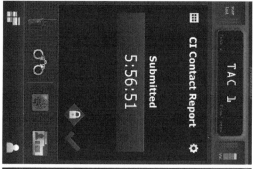

Figure 1. Conceptual handleld screens.

network surrounding a SWAT vehicle or a mobile command center, where many devices can take advantage of a vehicular modem to link back to headquarters. Handhelds also can help document an incident as it is occurring, with data sent directly to the network without the additional time or cost of manual reentry.

Whatever specific form LTE handheld units take, the goal is to put what's in the network in the hand. New handheld devices will be designed to deliver the best and most relevant information to first responders in real-time wherever they are—on the scene or en route—providing maximum situational awareness, so they can do their jobs more efficiently, more effectively and more safely.

The application revolution

Another enormous benefit of open standards-based networks is the variety of new public safety applications they will enable. It seems everybody is developing applications; that means lots of innovations are coming. Especially important are powerful multimedia applications that can combine voice, data, and video to power the benefits of collective intelligence. LTE networks will facilitate an ever-growing portfolio of public safety applications for enhanced incident management. A long list of innovative public safety applications includes CAD, analytics, still images and streaming video, facial recognition, next generation 9-1-1, access to large maps or data files, work ticket management, automated license plate recognition, and more. What is even more intriguing is the wealth of new applications we don't yet know about: those that are now in development or still gestating in the imaginations of developers.

A challenge that CIOs will face in harnessing this source of innovation is in first safely incubating and proving in an application without placing first responders at risk. Once the application is proven, it must be moved into a stable, trusted, and maintained framework for widespread use. The "application store" concept applied to public safety seems to be a reasonable method for distributing applications and managing them through their lifecycle.

Many applications for public safety rely upon access to highly secured but disparate government databases. In order to promote the development of applications by the widest possible community, publicly accessible mock-up or test databases will need to be provided for developers to use in creating and testing their applications. The data schema of these databases will obviously have to be published, as well. Finally, a procedure will be needed to certify the application, verify its integrity, and convert its use to live database sources in order to move the capability safely into the hands of first responders.

Streaming video

The first application most public safety professionals usually talk about is streaming video. Everyone wants to see what's happening, where it's taking place and who is involved, and video plays a huge role in making this all possible. "Video is becoming increasingly prevalent in almost everyone's life," says Schrier, "but it's mostly one-way video—us watching television or a movie."

In the collective intelligence era, public safety officers need to have two-way video to optimize real-time awareness of fast-evolving incidents. "The problem is most networks cannot support two-way video, other than in applications like Skype," adds Schrier. "LTE is transformational because for the first time, wireless two-way video becomes possible for public safety departments." One relatively recent example is when New Orleans first responders were directing the evacuation before Hurricane Katrina. Fixed and mobile video cameras distributed throughout the city contributed to the collective intelligence, showing which arteries were clogged and which could carry more traffic. These video images, streamed to command centers and back to vehicles, helped officers redirect traffic to make the evacuation faster and smoother.

Video is also used in a host of other public safety applications through its ability to provide cost-effective tiering of image quality. At basic scanning quality, you can remotely monitor for large problems, such as a bridge collapse or traffic tie-up. At a somewhat higher quality, video provides timely situational awareness, like identifying the number of people and vehicles involved in an incident, to help determine the most appropriate response. Even higher resolution images allow you to make better tactical decisions by helping manage on-scene resources, distributing Be on the Lookout (BOLO) images, identifying license plates, and much more. The highest resolution video helps agencies direct and deploy response teams in real time, positively identify suspects and assess the number of injuries, as

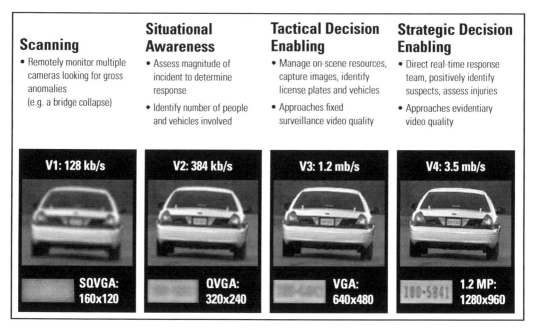

Figure 2. Simplified video quality chart.

well as provide evidentiary quality images. This kind of "augmented reality" helps officers be both better informed and safer. Obviously, higher video resolution requires more network bandwidth, so the balancing act is to intelligently adjust the video resolution in real time based upon factors such as the amount of bandwidth available, relative priority, and how it is being used by the collective for the situation at hand.

Mobile applications

First responders are major beneficiaries of innovative mobile applications. Today, rapid file transfer of large files allow the real-time downloading of mission-critical data, such as images of suspects or persons of interest, building floor plans, equipment manuals, database searches, and much more. Applications such as in-field reporting and mobile fingerprinting make the traffic ticketing process faster, easier, and more accurate, helping reduce the number of rejected tickets and the amount of lost revenue. Automated License Plate Recognition identifies a license plate, automatically matches it to a hot list, and immediately delivers crucial information on the car owner.

Mobile applications also are revolutionizing the EMT world. Western Fire Chiefs Association CEO Jeff Johnson says, "There was a TV series from the '70's called 'Emergency' that told the stories of medical technicians and how they used the technology of the day to save lives. Back then, it was basically communicating patient vitals by telephone to doctors while en route to the hospital. John Gage and Roy DeSoto would probably be appalled to find that it's not that much different today." But LTE and new mobile applications are now significantly improving on those decades-old techniques. "Today, especially in advanced life-support situations, technology can transmit not only full 12-lead EKG data to the receiving hospital, but also full critical-care patient parameters such as oxygen levels in the blood and more," continues Johnson. "They can also transmit full patient photos and videos, allowing doctors to "see what they see," helping them diagnose the patient and prepare for treatment even before the ambulance arrives." That definitely helps save lives.

Field-based applications also include remote monitoring of assets, such as bridges, roadways, and perimeters via embedded sensors and strategically placed video surveillance cameras. Public safety forces also use innovative robotic solutions such as units that can collect intelligence in dangerous locations, and bomb disposal robots that can be controlled remotely. Field-based reporting applications help officers stay on the streets longer with real-time electronic submission of incident reports instead of having to drive back to the station.

Next generation command and control

Perhaps the most basic, yet also most essential, next-generation application is CAD. Next-generation CAD streamlines the capture, correlation, and sharing of mission-critical information to optimize response, and increase officer and community safety. Some public safety operations are interested mainly in CAD or text-based dispatch, including next-generation 9-1-1, which is essentially concerned with bringing multimedia into a 9-1-1 call or event. The system ensures that first responders have all the mission-critical information they need exactly when and where they need it.

CAD systems make it easy to capture, correlate, and share the collective intelligence with first responders and other agencies and jurisdictions. They deliver mission-critical information from multiple sources on a single screen or console in real time, helping improve response and increasing efficiency and effectiveness. Many agencies also are considering next-generation 3-1-1 citizen response systems to handle non-emergency calls or requests, improving response and constituent satisfaction while freeing up the 9-1-1 system to handle emergency calls exclusively.

Second responders

Some jurisdictions prefer more complex systems that can integrate their entire community including local utilities, as well as integrate with regional, state, and federal law enforcement agencies, even the military. Some CIOs, including Seattle's Schrier, believe the LTE system also should include second responders. "I've been making the case for inclusion of second responders, utilities and transportation departments. I believe they ought to be on the LTE networks. When you've got an electrical outage in a whole neighborhood, the reality is your actual first responders will probably be the utility company. When we had an earthquake here in Seattle in 2001, we had cops and firefighters closing off bridges and highways when we should have had transportation workers there to close them off and inspect them for damage."

Analytics and collective intelligence

Analytics are becoming increasingly important in the collective intelligence world. Edge and infrastructure analytics help capture, correlate, and manage multimedia information in real time and for post-incident archiving and analysis. Analytics also can be proactive in alerting and supporting the command center when something unusual happens. They allow the extraction and sharing of the most meaningful information from a sea of available video, voice, and data. Analytics allow you to detect motion and perimeter incursion, isolate a face in a crowd, measure weather and environmental change, monitor traffic build-up, sense bridge or highway stress points, and a great deal more. By combining powerful analytic capabilities, automated network intelligence, and seamless interoperability, video solutions can deliver a single holistic view of each situation.

In addition to optimizing response and incident resolution, public safety departments also use captured data to analyze system-wide intelligence, using insights to take preventive measures that help improve community safety. In Fresno, California, for example, the police department uses its eCitation system and other mobile applications to capture and track demographics, such as race and sex, to help in defending against lawsuits. They also use the data to analyze traffic statistics and trends. Captain Andy Hall, head of the Fresno Police Department's Traffic Bureau says, "We can overlay data on Google maps that shows us where collisions are occurring and tickets are being given, which enables more targeted enforcement. We can look at those areas and make traffic engineering adjustments or increase ticketing activities to help reduce collisions."

The future is now

When we talk about innovation in public safety networks, devices and applications, and the new focus on collective intelligence, we're not talking about some time in the future. We're talking about now, as a number of systems are being planned and implemented around the country. Some of the most exciting projects that Motorola Solutions is involved in are the nation's first public safety LTE systems in Harris County, Texas, and the Bay Area Regional Interoperable Communications System, or BayRICS, which will serve multiple agencies across the greater San Francisco Bay Area. The BayRICS system is designed to integrate voice and data into a single public safety mission-critical network. It will consist of well over 100 700 MHz sites that are co-located with existing LMR sites. It is designed to permit a greater number of users to communicate using a variety of advanced information sharing tools. It also will support innovative daily applications—such as CAD, automated vehicle location (AVL), mapping services and records management (RMS)—as well as innovative new applications, such as biometrics.

The standards-based BayRICS system also will interoperate with other LTE public safety systems, such as the planned National Broadband Network, and can expand into regions throughout California and the West Coast. The Bay Area is working closely with the FCC, Emergency Responders Interoperability Center (ERIC), National Institute of Standards and Technology (NIST), and other groups to ensure interoperability with other systems.

Funding options

One of the biggest challenges public safety departments face as they enter the new era of collective intelligence and LTE is funding. This need not be an insurmountable hurdle, especially with the introduction of the FirstNet program and recent legislation that supports the reallocation of D-block spectrum to public safety. These programs and other funding initiatives over the next decade, supported by the president's administration and the U.S. Congress, will continue to fund the long-term implementation of a nationwide public safety broadband network. In addition, a number of existing federal grant programs also may be used to help local jurisdictions deploy this type of technology.

Beyond grants, there are innovative business models that can be investigated to implement this new technology. The first priority is to get these new life-saving tools into the hands of first responders, and allow them and our communities to reap the benefits. Flexibility can be the key. Secondary responders and other governmental agencies potentially could be given access to these systems on a secondary basis in order to help defray costs. We are already seeing unique, flexible business models in which some technology may be owned by the jurisdiction, while other technology may be owned by a vendor. This is another aspect of the public/private partnership that may, in certain circumstances, facilitate the deployment of public safety LTE networks.

Tough and timely questions

There are also a great many other considerations involved in planning for the collective intelligence era, which lead to a series of additional tough and timely questions:

- What does the future hold for our organization?

- How can we keep our network open, flexible, and secure?

- What range of applications will we offer? Can they support advanced technology such as video, mobile applications, data transfer, etc.?

- What are the various types of devices we should consider?

- What levels of prioritization and service level agreements (SLAs) do we need?

- How can we maximize leverage to all potential beneficiaries? How do we ensure the ability to work across agencies and jurisdictions?

- How can we forge a public/private network relationship that is both cost-effective and additive?

- How can we develop a network evolution plan that ensures the network, applications, and devices are kept current with industry advances?

By answering those questions thoroughly and honestly, CIOs and IT directors will be able to develop the right business models and the optimal 4G network scenarios while deploying the innovative crime-fighting collective intelligence tools of today and tomorrow. The result will be fewer unknowns, which will lead to more effective first and second responders, and safer communities.

PAUL STEINBERG is senior vice president and chief technology officer for Motorola Solutions. He oversees the development and execution of Motorola Solutions' technology strategy and vision, driving focused innovation around market needs and customer challenges to enable the future of public safety and enterprise. Throughout his 20-year career at Motorola, Paul has held several leadership positions, most recently serving as chief architect for integrated command and control and private broadband solutions for public safety systems. Prior to that role, he was chief architect for carrier wireless infrastructure broadband products (LTE) in Motorola's wireless networks business. Paul serves on technical advisory boards for multiple companies that supply products and technologies into the networking industry. He is a member of the FCC Technical Advisory Council and serves on the FCC's Technical Advisory Board for First Responder Interoperability. In addition, Paul is a member of Motorola Solutions' Science Advisory Board Associates (SABA) and holds several U.S. patents. He was recognized with the Motorola Dan Noble Fellow award in 2004 and named a Motorola Fellow in 2006. Prior to joining Motorola, Paul was a distinguished member of technical staff at AT&T Bell Laboratories.

Chapter 3: Successful Technology Development in Public Safety

BY LEONARD SCOTT

Delivering on technology needs in public safety can be scary or seemingly impossible. Just look around and see who has "succeeded" and who has failed in this effort. Sometimes both ends of this continuum are seated with the same technology from the same vendor. How can things be so divergent? Why does one technology project go forward as a great success while the same technology fails when installed in a neighboring agency? A peer and mentor tells you, "Whatever you do, don't buy brand X. Take a look at what happened to Metropolis when it tried to use that system: It lost millions, and our friend Bob lost his job as a result of the failure." Or, "Take a look at Cyber City and its system Y. It does everything they need, and they got funding to cover it all."

One of the quotes heard quite often is, "We have applied over and over again for grants and have been turned down every time. My agency does not receive funding that will cover the expense of this crucial technology." Or, "My council supports/funds most other government projects but sidesteps public safety projects because there is so much federal money available for them." How about, "My radio system no longer meets our needs, and funding is not available for me to move to a good trunked public safety radio system. Those guys over at public works are getting their system replaced. What gives?"

Experiencing success in the implementation of public safety technology requires much more than shopping and buying a system advertised or installed at another location. Quite often, agencies will spend more time and money in the selection and implementation of a fleet of vehicles than selecting a voice communications system or a computer aided dispatch system. Selection and implementation of technology systems is not fool-proof and requires much thought, preparation, and planning coupled with professional project management to better ensure success.

Planning

The primary key to success is planning. An agency that properly plans and shares its plans with those responsible for budgeting positions itself to receive funding and support where it counts. Planning should include short-term, medium-term, and long-term documents. Anything less is not complete and stands to lose the support of those who decide what is important and what will gain the most for the agency in terms of efficiency and safety.

A short-term plan is generally nothing more than watching for required changes that have taken place after the establishment of the current budget and the immediate actions that must be taken to comply with these requirements. This phase of planning "plugs the leaks" in current operations and adjusts the requirements for both medium- and long-range plans. Not only should short-term plans account for changes in laws and public expectations, but also what is happening with current technology availability and upgrade opportunities. Anything less than the best can spell disaster for public safety operations. The public expects the best and feels that they have paid for that level of protection from their public safety providers. An avoidable death or massive loss of property can spell disaster for elected/appointed officials if the media finds that available technology could have averted the loss and at some point in the planning process the technology was skipped or missed. The key here is to keep abreast of the ever-changing status of technology, and plan for it to be addressed.

Medium-term planning is for the next three years of operation in technology and support. Medium-term planning includes acquiring budget support and planning for the implementation of new or improved technologies. Needs assessments must be completed to articulate what the department lacks, and a needs assessment must be converted to a "functional" Request for Information (RFI), Request for Proposals (RFP), or similar documents to quantify the agency's need for the technology, and what it can provide in terms of cost savings and/or safety. These documents should portray to the approvers your effort to quantify the cost of the technology in terms of cost savings or cost avoidance. Safety issues should be explained. The document should contain (among many items) a disclaimer allowing the agency to refuse all vendor solutions/proposals. This allows the final decision for implementation to be in the hands of the decision makers.

Long-term planning is for four years and beyond. At this level of planning, all existing technology should be reviewed for support availability, available upgrades, market changes, and potential superior solutions. It is important to keep up to date with technology to keep from falling behind market availability and public expectations.

Review technology that is in development and its potential availability. Don't get caught behind the curve, as this identification, education, and development is your responsibility. Don't get caught sitting on your hands by an appointed or elected official. Quite often, vendors begin early educating your superiors about upcoming technologies before they approach you. Their intent here is to make your job easier, so it is best you keep abreast.

Be sure to keep your RFIs and RFPs "functional" in nature to stay brand/technology generic. You should describe carefully what you want for the technology to do for you at every level. Leave the engineering work to the experts, and allow them the "technical license" to develop/propose their best solution. Make sure that your evaluation team objectively evaluates all qualified responses to select the best response in the confines of your budget.

Where will the money come from?

One of the most important aspects of the funding process is to research potential partners to share the benefit/cost of technology. Potential partners are diverse and abundant.

Communication systems, location systems, social network interfaces, etc. are technologies that can be shared by many government agencies and private businesses. Don't exclude anyone by thinking that we can't share public safety systems with others. Technology has long been in place to protect public safety operations in these sharing scenarios. Being able to take advantage of "economy of scale" can bring in valuable dollars and enhance the acceptability of your efforts. Some technology, like a transit/public safety partnership, can bring enhanced benefits to the table. Quite often in incident command situations, there is a need for quick evacuation, and shared technology can enhance our ability to provide the public with time-saving enhanced operations. The same is true with any partnership. Remember on 9-11, inter-agency communication and interoperation were the needed links.

Grant providers are very interested in the potential impact of their funds. If your proposal will serve to enhance your agency only, you will quite often be placed behind a multiagency proposal that enhances public safety as well as other entities.

One thing to remember is that we do not exist on grants alone. Although we should look to share our opportunities, it just might not be possible politically. Remember that agency budgets are based on projected funds divided by existing needs. If we can have a successful impact on existing needs by placing a needed unfunded project on the table, the decision makers can elect to adjust overall funding to support an essential technology. Money can be borrowed over a period of time to reduce the impact on annual budgets. Consider asking to have one of your elected officials serve on your project committee from the needs assessment all the way through implementation and acceptance. If chosen wisely, this official will assist you in the political arena by helping their peers understand the project and its potential impact on their constituents. This level of understanding and support can be invaluable to the eventual success of your project. Money exists in the overall budgeting process as long as decision makers understand the overall importance and potential impact on the community.

Who should help with the project?

The members of your project committee are very important in the overall potential for success. It is important to have representation from IT services. It is also very important to have representation from potential users. This representation should include all agencies to be served by the technology, including unions. The overall balance of your committee is important to ensure that all the needs possible can be identified and addressed. This will help with rumor control.

Some members of your organization may find an opportunity to secure attention by starting rumors based partially on truths, which draw attention away from the project and serve to minimize its true importance. Such diversions can serve to remove the proper support for funding. It is important to address all rumors at their first appearance by evaluating their validity and then addressing any resolution or by making public the false assumption along with materials that show that the concern is not valid. It is important to display to all that there are no ulterior or secret motives for your project and that the project appreciates the expression of any concerns and promptly addresses them publicly.

Reduce your load to give more time and resources to your project

As the person responsible for daily operations of your organization, you often don't have the time necessary to devote to special projects. The responsibility to replace technology, wrestle funding for operations, and fight to follow disposal responsibilities for equipment that has no current value can eat up anyone's resources. By streamlining this process, an organization can free resources necessary for daily operations as well as for project management.

Many agencies consider technology-related services as capital items. In the period of the 1950s through the early '70s, this was the correct identification. With the advent of the personal computer and VoIP phone systems, to name a few, the situation has changed. Items that before were major purchases are now a necessity of daily operations and should be considered an expense by your organization. Our dependence on these items is no different than our need for electricity, water, etc.

With this in mind, consider a lease replacement program for current equipment where a successful bidder can supply systems based on declared requirements with a two- or three-year turnover at an established monthly charge. At the end of the two/three-year period, systems are replaced by the supplier with all information on the old system transferred to the new. Old systems are then removed by the supplier. Maintenance and support also should be included.

This change will ensure proper identification of monthly/annual costs for technology and require commitment from user organizations regarding their requests. If I want a PC, it will cost me a known amount, or to add a phone instrument and line, I can identify the annual cost. End users are assured that they always will have new equipment and support.

What does the future hold?

New demands are surfacing from a public that has more instant information at their fingertips supplied by PDAs and computer tablets. It is difficult for this informed public to understand why agencies are not prepared to accept their assistance when it comes to photographs, videos, and audio recordings they have obtained and want to make available to first responders. 9-1-1 agencies and first responders are struggling with this issue, and help is on the way. Unfortunately, funding is not always identified, and support by elected officials is not always present. Yet, this information quite often can make the difference between solving a crime or even the difference between life and death. The technology is referred to as Next Generation 9-1-1 or NG9-1-1.

What might follow NG9-1-1?

The same informed public asking for NG9-1-1 is wondering why public safety can't inform them of events that might have a direct impact on their daily lives. Think of how Facebook works. It interacts with you based on who you know and what they are doing. Facebook can alert you to things that directly impact you, and the same can be true regarding issues relating to first responder activity. If we were to complete a confidential profile of activity in a social network (non-existent yet) regarding my family, where we go and how we get there, it would be relatively easy for a computer-aided dispatching system to send me information regarding calls that could impact my family. Not only could I receive this information, I can choose how I want to receive it, i.e., text message, voice message, fax, email, etc. Thanks to recent developments in technology, I can receive this information in most any language I choose. We can translate most any spoken language into text and then change it as many times as necessary for dissemination. I also can switch this text into any spoken language. This interesting revelation has many applications, as suddenly the ever-evasive science fiction "translator" becomes a reality.

What about follow up by the public or case updates by investigators? It is forever the case that when an investigator contacts a complainant regarding a filed complaint, that the victim has additional information that should be used to close or update the case. This is true with many missing persons and stolen vehicle cases, in particular. Being able to exchange information through a social network allows us quick access, and victims can update authorities, and authorities update victims, allowing official information to be more complete and reliable. This is a pleasant update to the current status of authorities chasing after victims by phone and never making contact, or victims not really knowing how to contact authorities regarding changes in the status of cases filed.

I am reminded of the case of a stolen vehicle whose owners (wife retained her maiden name) reported their vehicle stolen to local authorities. The vehicle was registered in the husband's name. The couple recovered their own vehicle the next day after they discovered it had been borrowed by a relative in their absence. They were not sure what to do about the theft report, so they decided to check it out "later." Well, later became never, and when the wife was on the road two months later to visit relatives in another state, she was stopped and arrested for auto theft. It was several hours later that authorities were able to contact the husband, who obtained her release.

This is one case in many that can be averted only by agencies that keep up with current technology and keep their employees trained and up-to-date on its use. The better job we do in this regard, the better, safer, and more crime-free lifestyle we will lead.

LEONARD B. SCOTT is a retired Captain from the Corpus Christi, Texas, Police Department, having 36 years of distinguished service. At retirement, his service to the department was the Command of the department's technology operations, including planning for, as well as recovery from natural and/or manmade disasters. In February 2006, Mr. Scott was named by Government Technology Magazine as one of the "Top 25 Doers, Dreamers and Drivers in the United States." He has served as a primary speaker at events throughout the world and continues to be quoted by leading publications as a primary strategist for wireless data systems and software. In addition to serving as a full-time instructor for Coastalbend College, Mr. Scott serves as the Director of Public Safety Technology Programs for the Public Technology Institute (www.pti.org).

Chapter 4: Public Safety IT Budgeting

BY JEFF WEBSTER

The Obama administration's FY 2011-2012 budgets include significant cuts to funding for Department of Justice, Office of Justice Programs, dictating that state and local agencies will be forced to think ever more creatively when it comes to IT budgeting. This sentiment is felt more in the interoperable communications world than any other public safety realm. Given tight mandates and deadlines from the FCC, agencies are scraping for additional funds to meet the narrowbanding deadline of January 1, 2013. Public safety agencies are fortunate in that their sense of urgency for complying with these mandates is usually met with funding, as seen throughout the past two administrations. However, 2012 and beyond looks more bleak than previous years for federal funding. Agencies will need to promote the need for public safety more so than ever in the coming years in order to obtain necessary federal and state funding. Fortunately, public safety has had a strong political track record, and many city, county, and state political figures tend to avoid cutting spending on public safety based on its political ramifications. This chapter will examine the process of budgeting for public safety projects, specifically interoperable communications, and the procurement process that follows.

Budgeting

When it comes to implementing new technology infrastructure, the hardest question for a CIO to answer is always, "How will we pay for this?" The response to this question usually is followed up by another question: "What are our options?" This is the point in which agencies differ on their response. The decisions an agency makes on its funding sources can steer a project down a path of success or a path of destruction. No matter what decisions are made, the impacts of these decisions are felt across the agency. These funding decisions have the ability to allow the organization to continue operating as normal, or it can put strain on the organization. It is imperative that CIOs equip themselves with the tools necessary to ensure fiscally sound decisions are made.

CIO Budgeting Tool Kit:

- A comprehensive and sound budgeting/accounting/reporting program
- Educated and trained staff
- A team of thought leaders from different stakeholders
- A repository of historical data
- Grant resources

Sound budgeting/accounting/reporting program

The value of a comprehensive and sound budgeting, accounting, and reporting program cannot be overstated. In a world where "going green" usually means reducing your carbon footprint, "going green" to a CIO means streamlining and reducing redundancies. It is through a comprehensive and sound budgeting, accounting, and reporting program that workflows are streamlined and efficiencies are identified. It is also through this program that CIOs have the ability to work faster, smarter, and see results in real-time. Not only is the CIO smarter, but the decisions he/she makes are more data driven as opposed to emotionally driven. Removing emotions from the budgeting process can improve success and build credibility among the organization. Too many times are there examples of line items in a budget that are backed by an "I know what is best" mentality versus a "This is what the data says" mentality. Most of these examples stem from a lack of proper accounting and reporting. This opens the doors for misconceptions or pre-conceived assumptions on what should be included and at what level.

Assumptions and pre-conceived notions should be avoided at all cost when budgeting for public safety communication purchases, upgrades, or maintenance. When looking at these three budget line items, chances are a CIO is making decisions on items that have a potential spending level of twice your annual personnel budget. And, depending on the size of the organization, this number could be three, four, and five times the size of your highest annual expenditure. This means you will experience heavy oversight and scrutiny on each proposed line item. Therefore, it is imperative that you equip yourself with the necessary tools to curtail any concerns that may arise when seeking budget approval.

Not only will these tools help curtail your toughest opponents, but they also will be the tools you need to identify gaps and evaluate consolidation and resource reallocation options. Given the size and scope of public safety communication projects, it may become necessary to look at your current budget to help identify new funding options. This includes looking for spending and expenditure gaps. Such gaps can open many doors for new projects or programs. By equipping yourself with a tool that can report on revenues versus expenditures, you are more likely to identify these gaps early, which in turn allows for more precise budget decisions. An example would be in an agency's 911 surcharge fund. Knowing how much revenue is being brought in and where that money is going could lead to more informed decisions on how best to use those funds.

Consolidation and resource reallocation are becoming commonplace as state and local governments are experiencing the worst financial environment in years. City councils, county boards, and state commissions are forcing agencies to cut costs and do more with less. Public safety organizations are continually searching for ways to free up funds to pay for new purchases and costly upgrades. Thus, many agencies travel down the path of consolidation or resource reallocation to free up funds. Understanding your current financial environment makes consolidation and resource reallocation a lot less painful.

Educated and trained staff

This tool is especially important to a CIO because nine times out of 10, the CIO is making budget decisions from the recommendation of his/her staff. An uneducated and under-trained staff will lead to budgetary blunders and miscalculated decisions. In order to mitigate this, it is important for the CIO to build a team of well-educated and trained professionals who have the experience necessary to make sound recommendations. In order to do that, a CIO should build a team consisting of the following:

- **Number crunchers**—These are people that know your budget inside and out. They are also the ones that can accurately predict expenses and revenues.

- **Statisticians**—These are people who can sift through years of budget data to identify trends or commonalities. They also can look at economic trends in your jurisdiction or the private sector to better understand costs.

- **Technical experts**—These are people who have the proper IT knowledge to understand exactly what type of expenditures are needed for IT line items. These people are closely aligned with other technical staff and should have knowledge of existing equipment and infrastructure.

- **Innovative thinkers**—These are people who can look at your budget in a different way and identify new areas for revenue or funding. When evaluating consolidation, these are the people to rely upon.

A team of thought leaders from different stakeholders

No matter what type of technology you are implementing, there is going to be a core group of people who have vested interest in the process and outcome of the project. Identifying who these individuals are and engaging them early in the budget process will help avoid confusion and project run over. Typically, these individuals are department heads or upper management, but may include mid-level technical folks. It is important to have a good mix of individuals representing a variety of aspects within the project. These aspects typically involve operations, management, leadership, development, and training. Engaging individual(s) who are responsible for these core functions will ensure that your budget is clear, concise, and complete.

When working with these stakeholders, it is important to hold each person accountable for their budgetary responsibilities. These responsibilities could include current budgetary analysis, technical environment, and operational management. By having stakeholders with vested interest in a project or annual budget, you are more likely to complete the budgeting process on time and within scope. However, should your stakeholder group include private sector assistance, you will want to make sure that information provided to the group is accurate and backed in writing. There have been far too many examples of public safety agencies receiving budget figures from private sector partners that are nothing more than an educated guess. This will cause major headaches further down the process and could result in a cancelled or severely delayed project (i.e., New York's Statewide Wireless Network and Oregon Wireless Information Network).

A repository of historical data

Having historical data within arm's reach can be essential in making sound budgetary decisions. Historical data will not only guide a CIO during the budget process, but will also be a key driver in defining goals and objectives for the next 5+ years. Valuable historical data includes revenue and expenditure data (month/quarter/year), population data, technical environment data (equipment and services), and grant data (state and federal). While there are plenty of other data sets that must be considered during the budgetary process, these four should guide a CIO in the right direction.

When planning for and implementing a new technology or project, the most obvious piece of information needed is past, current, and future revenue and expenditure data. These data sets are the key drivers in determining the feasibility of a new technology or project. Simply stated, if you don't have the funds or cannot acquire them, you probably should not pursue the project. However, there are ways of planning for future funding streams. The key take away is that revenue and expenditure data will be the key driver for any project budget. This is where a CIO should lean on his/her statistician to uncover trends or commonalities among previous year spending/revenue.

Other useful pieces of historical data include population and technical environment data. Both of these data sets can be the catalyst for any new technology project, ranging from communications and dispatch to records and information sharing. For example, using population data when budgeting for a project can give insight into how much strain will be put on a system. A communications system used in a town of 50,000 people will see less strain than if the town grows to 75,000. Knowing population trends can assist an organization in properly budgeting for upgrades and maintenance to any communication system. The same can be said about technical data. Inventorying and identifying end-of-life schedules can be critical when trying to budget three to five years past the original purchase date. On top of that, knowing how long a vendor will support equipment is also helpful. Given that public safety is a product-based market, these products don't last forever. Radio communication equipment is especially prone to expiration. Organizations need to properly plan and budget for technical improvements 5+ years in advance to any expiration dates.

Grant resources

As most agencies find, there is not enough money within a budget to cover the cost of most technology projects. In order to make up for this gap, many agencies turn to the federal government for supplemental funds. This is common practice for most public safety agencies who are seeking to upgrade existing infrastructure or implement innovative policing technologies. Below is a flow chart of how a typical grant program works:

Figure 1.

The process of applying for a grant and winning funds can be a long and strenuous process that takes multiple years. In order to ensure success within your agency, the following resources should be acquired before any plans are made to pursue grant funding:

- **Access to www.grants.gov**—this website is run by the federal government and allows agencies to view, understand, and apply for grant programs. It will be the best resource for information on latest news, updates, deadline modifications, and application requirements.

- **Access to previous grant applications**—this resource will be most valuable when deciding what to include or omit from your application. The Department of Justice, Bureau of Justice Assistance has a variety of resources for agencies looking to apply for grants. They offer access to previously awarded grant applications, which will be helpful in determining what your application should look like.

- **Grants management system and/or team**—nothing is more important than editing and meeting deadlines. It is commonly known that grant applications are automatically disqualified should the application not meet defined specifications or be submitted late. Using a management tool or building a grants team will aid in the process and ensure that your application meets all specifications and deadlines. Any grant team also should consist of number crunchers, statisticians, and stakeholders to ensure that the application is unique but specific to your organizational goals/needs.

On top of knowing how the grant process works and what resources are needed to produce a winning application, it is also important to know what grant programs offer the best chance for funds. The table on the following page outlines commonly used Department of Justice and Department of Homeland Security IT grant programs with approved budget figures. There is no guarantee that these programs will be in place in subsequent years and as priorities change, but an understanding of these programs are key to any CIO seeking public safety funding.

IT procurement

When looking to implement a new technology or purchase new equipment and services, a CIO will most likely encounter roadblocks along the way. However, those roadblocks vary

Table 1.

Common Public Safety IT Grant Programs (approved funds $M)						
	FY 2008	FY 2009	FY 2010	ARRA	FY 2011	FY 2012*
Byrne/Justice Assistance Grant (JAG)	$374	$755	$744	$2,225	$430	$519
Community Oriented Policing Services (COPS)	$587	$582	$791	$1,000	$537	$669
Urban Area Security Initiative (UASI)	$782	$799	$833	$0	$725	$920
State Homeland Security Grant Program (SHSGP)	$861	$861	$842	$0	$725	$1,050
Port Security Grant Program	$389	$389	$288	$160	$235	$300
Assistance to Firefighters/ SAFER	$750	$775	$810	$210	$810	$810

*Budget recommendation

in size and shape. Some roadblocks are easy to overcome, and some are more difficult. But, one thing is certain: no roadblock can be as insurmountable as the roadblocks that a CIO may encounter during the procurement process. In order to avoid these scenarios, CIOs need to learn the proper way of preparing for, handling, and finalizing the procurement process. A lack of planning and understanding can bring a project to a halt or even cancel it for good. The typical procurement process is as follows:

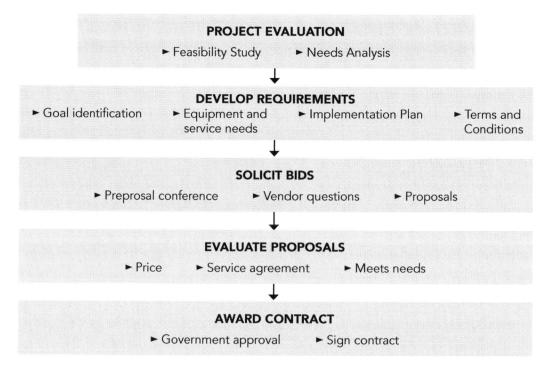

The procurement process does not start with a requirements document. Effectively procuring the right equipment and services involves in-depth planning and analysis prior to crafting a solicitation document. This could not be truer than in public safety. Given the dynamics of the market and the technologies it has to offer, public safety IT purchases involve multiple facets of an organization. CIOs should begin the procurement process by outlining goals and projected outcomes. During this stage, it may be necessary for an organization to seek a third-party consultant to help with:

- Inventorying
- Feasibility studies
- Needs analysis
- Requirement documentation
- Proposal analysis
- Implementation strategies

Some agencies have the capability to complete this work on their own, but most agencies do not possess the necessary knowledge base to fully understand what the organization's needs are and how to obtain them. The simple fact is that this is probably the first time the CIO has implemented this type of project, whereas third-party consultants perform this work daily. Not only can a consultant ensure complete and robust requirements analysis, but they have the ability to save the organization a great deal of money. This cost-saving aspect will come in handy when proposing a project to the city council or county board.

Once an organization has gone through the process of evaluating existing infrastructure versus planned infrastructure, the next step is to draft the requirements document. This document will provide vendors with the information needed to submit proposals for the work requested. When drafting this document, it is important to be as thorough as possible in order to avoid confusion among potential bidders. This confusion can lead to higher proposal pricing or unnecessary equipment/services. It is during this stage that a CIO should rely heavily on their technical experts (consultants or in-house staff). These people should have a firm grasp on the goals and outcomes of the project in order to better understand what requirements are needed.

Also included in the requirements document should be a scorecard. This scorecard will outline what the agency needs and what percent these needs weigh on others. This scorecard will give vendors a clear outline of exactly what the agency is looking for and how much value each component has. This scorecard also will be very valuable to an agency's proposal review team. It allows for quick and accurate scoring of proposals and can be a good piece of collateral to bring to any city or county board meeting.

After the requirements have been drafted and the organization is ready to release the solicitation package, it is best to leverage as many tools as possible to inform potential bidders of the solicitation. These tools include city/county/state procurement websites, local newspapers, industry associations, etc. When seeking contract approval from the city, county, or state, it will be beneficial to outline the competitive nature of the procurement. Almost all city, county, and state governments require or strongly promote fair and open competition for any procurement. This typically applies to procurements exceeding the pre-determined threshold for projects requiring public bidding. This has more to do with wasteful spending than anything else. By showing best value, a CIO can avoid additional roadblocks that might have the potential of either cancelling or forcing a re-bidding of the project.

The next step in the process is to release the solicitation. This step will include a vendor pre-proposal conference, vendor questions and answers, and additional addendums or amendments to the solicitation. During the procurement process, it is important for a CIO to understand his or her limits when it comes to vendor outreach. Knowing who is allowed to respond to vendor inquiries and what type of responses are required is crucial. This is important because anything said or given to a single vendor could be construed as unfair. This has the potential to result in protests by unsuccessful vendors, which in turn prolongs the procurement process and costs the organization in both time and money.

Once proposals have been received, it may be necessary to develop a proposal review team. This team would be responsible for evaluating bids and scoring. They also would be responsible for making recommendations based on scoring and evaluations. These members should be made up of a variety of different stakeholders from the user and technical agency to the administering and financing agency. Having a robust review team ensures a fair and comprehensive review process that leaves no room for protest. This team also will be a good resource to have during council and committee contract review meetings.

The goal of the proposal review process is to evaluate cost, timelines, needs versus wants, and implementation strategy. Cost containment is the number one priority of many project managers and CIOs. Therefore, it is imperative that a review team thoroughly analyze and review each vendor proposal. This includes every equipment cost, service fee, delivery fee, installation fee, and maintenance cost. Also, if this project is the basis for a grant application, it will be even more necessary to evaluate all costs as to not discredit your grant application. With new federal reporting requirements, agencies are held more accountable for spending their grant dollars in line with their original proposal. This sentiment is true for the timelines and implementation strategies outlined by the vendor. Not only do they provide insight into the probability of achieving project goals and outcomes, but they will be the backbone of many grant applications.

The evaluation team also should consult with other agencies during this process to gain insight into vendor performance. As stated before, meeting implementation deadlines and keeping costs to a minimum are essential. Reaching out to neighboring jurisdictions to ask about a vendor's performance can help justify a decision. There is a very strong chance that the technology being procured has been procured before. CIOs should build relationships with neighboring jurisdictions, as this will aid in the process today, tomorrow, and years to come.

The last stage in the procurement process is contracting. This stage is actually more the beginning than the end for a project. It is the point in which pavement meets the road. Once a contract has been signed, the project begins. In order to get to this point, an agency and vendor must agree to a formal contract binding each party to their respective requirements. Once the evaluation team has made their recommendation, the decision is typically sent to the council, board, or commission for final approval. Assuming there are no hiccups along the way, the approval process will conclude, and a contract will be executed. This ends the procurement process.

JEFF WEBSTER is a Manager at Deltek, focusing solely on the state and local Justice/Public Safety and Homeland Security Information Technology market. His team is dedicated to providing clients with the intelligence needed to win more government business in the areas of public safety communications, 911, information sharing, courts, and corrections. Webster has been featured in a variety of industry publications related to public safety communications, and he speaks at industry events across the country. Prior to joining Deltek, Webster worked for the Minnesota State Legislature and the Council of State Governments (CSG).

Chapter 5: Collaboration Technology in Public Safety Communications

BY PAUL WORMELI

Introduction

The purpose to which technology is applied drives its features and functions, and sets the limits and boundaries for what is included or excluded in product or service packaging. It follows, then, that it is helpful to define a purpose to which a set of technologies is applied in a single and very clear statement of such purpose. If so, then there is a value to defining a class of existing or emerging technology that is or will be applied to facilitate collaboration within and across communities of interest or domains. By having articulated such a class, we can define in more detail its attributes and objectives in ways that will encourage innovation and will allow us to re-purpose existing technology to apply it to this more specific goal (purpose).

The travails of the economy and the emergence of the networked world have made collaboration the order of the day, so much so that executives in either government or industry spend a considerable amount of their creative time trying to figure out how to be more effective through collaborating with their peers and even their competitors. If you want to see how far-reaching a significant this relatively new awareness is, just go to Mother Google and do a search on the phrase "Collaborate or Die." Lest you think that this title is overly dramatic, you'll be amazed at how many different fields now have this slogan on their front page.

The most articulate and powerful set of lessons learned as they apply to the public safety and justice field is found in a new book. Bill Bratton, former Chief of Police in Los Angeles and Commissioner of Police in New York, and now Chairman of Kroll Industries, and Zachary Tumin, a senior researcher and analyst at the Harvard Kennedy School of Government where he served as Executive Director, Leadership for a Networked World Program, have collaborated (yes, I used this word intentionally) on a book entitled *Collaborate or Perish!*,[1] which reveals the best practices that these two coherent and capable government leaders have developed to carry out this mandate.

This kind of a culture change is only sustainable if investments in the supporting infrastructure are designed to keep it alive. Training, a reward or incentive system, and supporting services are needed to make any culture change permanent. Collaboration is no different, and therefore, we should look at what kinds of technology can and will support this movement that has become so important.

Social networking tools such as LinkedIn, Yahoo groups, and wikis in general can be viewed as collaboration technology, but there is still a need to move these technologies forward to better facilitate collaboration through organizing the presentation of collective intelligence. The development of Next Generation 911 is in a sense collaboration technology in that it facilitates the multi-media exchange of information between citizens and public safety agencies. There is still a great need for innovation in our striving to apply "the wisdom of the crowds"[2] to decision making, and this is what collaboration technology should be all about. If we can better define the requirements for collaboration technology, assess the gaps in what already exists, and define the functions that future innovative offerings could provide, we will move faster toward making collaboration the rule rather than the exception. This chapter explores these recent technologies and other national programs intended to foster improved collaboration across jurisdictional and disciplinary boundaries.

Collaboration technology in public safety

There is actually a long history of collaboration technologies, dating back to perhaps 1951 when Doug Engelbart apparently first articulated the idea of collaborative software. The early versions of such products were called "groupware" and had the purpose of facilitating online meetings to share information. The origins of this genre are described in Wikipedia.[3] Today, there are many other categories of technology that have application in public safety, and we can define several different groups of these technologies available to the communications director:

Collaborating with the public

Historically, the most common technology used to communicate with the public has been the telephone. PSAPs often are compared to call centers in the commercial world, although there is no comparison in terms of the life-or-death mission of public safety communications centers. Nevertheless, the roles, architecture, and processes in a dispatch center are about answering the phone with either separate or consolidated call takers and dispatchers at work.

In these hard economic times, however, many people are searching for alternative processes and approaches to this interaction with the public. In Europe and the UK, there is an increasing tendency to view the relationship with the public in a way that makes citizens equal partners in the co-production of public safety. Some of this has been a feature of community-based policing programs for years, but this new idea of co-production sets a new and more serious tone for engaging the community. It also requires new technologies to support a different and extended role. Some of this is found in the rationale behind the development of the Next Generation of 9-1-1 as fundamentally an IP-based technology. NG 9 1-1 at least opens the door to the use of multi-media technologies to interact with the public.

With the proliferation of smart phones, most of which have good cameras, video recording, and both text and e-mail facilities, both victim and witness interaction often will offer a much more robust mechanism for communications and collaboration than just a telephone voice capability. People on the scene can be thought of as "sensors," offering immediate insight into the nature of an event and helping the dispatch center make better resource allocation decisions. Perhaps more significantly, residents (particularly the digital native generation) just assume that the public safety agencies will be able to receive at least e-mail and text messages, and they often are shocked when this is not the case. If we think about all these technologies in terms of how they contribute to or play the role of collaboration technologies, we can draw a better picture of what must be done to incorporate these capabilities into the communications center. They are perhaps the most pressing argument for moving to NG 9-1-1. Digital natives are notorious for relying on texting for communications, and there seems to be no better way to exchange information with this age group.

It may seem to be a stretch, but it is useful to think of video surveillance as a mechanism to interact with the public. There is a collaboration required particularly when the communications center is able to call upon the video cameras placed into service by businesses and the transportation department. The Atlanta police negotiated an agreement with major companies to link their street-based video cameras into the communications center to give the police the eyes of a lot more devices throughout the city than the police department could afford.

Collaboration is a two-way street, and there is much more that can be done to use technology to foster interaction with the public than agencies typically use. Today's low cost for providing webinars makes it possible for agencies to use this technology to provide seminars on crime prevention, target hardening, environmental protections, etc. Social media, such as Twitter, that provide for outgoing (push) messaging to subscribers can be used for making crime notifications to neighborhoods and shopping centers, alerting the community to searches for missing persons or animals, potential traffic disruptions, etc. Wiki software can be used to create working groups with, for example, merchants in a shopping center, to share ideas with the police on crime reduction as well as crime suspect descriptions.

Finally, in these days of limited budgets and fewer resources to assign to handle calls, forward-thinking communications leaders are working on ways to handle calls for service without dispatching resources. This is absolutely a collaborative effort with the responding agencies (police, fire, and EMS all can work on this). More and more cities and towns are seeing the value of creating a web-enabled crime reporting capability so that "let your fingers do the walking" takes on new meaning. For those calls for service that are reporting the type of crime that is really reported for insurance purposes, it makes no sense to send a highly trained police officer to respond when a resident can enter the crime report on the Internet. The online entry of less serious crime reports could save as much as one-third of the time spent by patrol responding to these calls. If the online reporting is coupled with the quick review and necessary expansion of the data so that notifications can quickly and easily be made, for example to the state and NCIC repositories, the probability of recovery rises.

Calls for service team collaboration

If we stand back and reflect on what is happening in a communications center, it is helpful to view the team involved in the response as a collaboration team. The call taker, dispatcher, and responding resources are all working together (collaborating) in the delivery of customer service. When a patient is involved, the team expands to include hospital personnel and physicians. Taking this view of the process, we can explore the question of what collaboration technology maximizes the likelihood of collaboration success. It is obvious that the process of responding to a call for service is intricately linked to information sharing. The call taker informs the dispatcher, the dispatcher informs the responding units, and the responding units report on the disposition. For this process to be effective, the computer-aided dispatch (CAD) system or its manual equivalent has to support the collaborative efforts, and so we should think of CAD software as part of the collaboration technology. However, there are other support functions that can be envisioned where technology can help ensure the best response. Web-based data sources available to all of the team so they may apply the best wisdom about emergency protocols and many other informational topics become part of the arsenal of collaboration software. Easy-to-use, non-duplicative data sharing becomes one of the keys to successful collaboration in a response.

Interagency collaboration

The procedures developed for mutual aid between agencies have long established the practice of collaboration between first responders to help each other in routine calls as well as major disasters. In communications centers, the organization of primary and secondary PSAPs also establish a formal and effective collaboration. There certainly has been collaboration technology supporting these agreements to provide calls transfer, routing, etc. However, the newer forms of collaboration have yet to see the support for information sharing to match the call handling in an analog world.

A significantly increasing capacity for information sharing focused on the exchange of data is now emerging thanks to the availability of standards like the National Information Exchange Model (NIEM). Now that the justice and public safety world have begun to implement NIEM, it is much easier to build information exchanges that can be passed between jurisdictions and communications centers. One of the first applications of NIEM was in support of a joint project sponsored by APCO and the IJIS Institute that focused on the exchange of alarm information between alarm companies and communications centers. In an effort to create a single standard for this function, a NIEM conformant message was designed and implemented on a trial basis in Richmond, Virginia. Once it was successful, APCO used its ANSI standards process to adopt a communications standard that any communications center can implement by referencing the APCO standard for alarm messaging.[4]

The implementation of NG 9-1-1 will have a significant impact on interagency collaboration. As NG 9-1-1 networks begin replacing the existing narrowband, circuit-switched 9-1-1 networks that carry only voice and very limited data, the IP-based NG9-1-1 network will be providing information such as text messages, images, video, support for hearing and speech impaired users, telematics data, and links to related resources like building plans and medical information to emergency communications centers, e.g., Public Safety Answering Points (PSAPs). All of the data submitted in reference to a call will need to be shared among the various responding agencies and, in some cases, multiple communications centers. There is a great deal of work yet to be done to develop standards, probably based on NIEM, to facilitate this kind of data sharing among communications hubs.

Achievement of a fully featured, truly standards-based NG9-1-1 system will require a focused, managed, and collaborative approach to the development of data standards, which must include standards for data exchange. One of these vitally important standards is a data exchange to transfer the NG9-1-1 data from the primary PSAPs to other PSAPs and other relevant public safety agencies and domains. The NG9-1-1 project's Emergency Incident Data Document (EIDD) Work Group has been working for more than 18 months to define the content, parameters, and use of the EIDD, and is now ready for their model to be formalized into a data exchange standard.[5]

Communications centers are getting used to new avenues of collaboration beyond the conventional wisdom. APCO has engaged with the program manager for the National Suspicious Activity Reporting Initiative in a program to train communications personnel to be aware of information coming from the community that could potentially have a nexus to terrorism. The center personnel are being trained on the various precursors that have been identified as being potential indicators of terrorist behaviors, and on how to report the information to fusion centers and task forces who have the responsibility to validate the information and take appropriate action. A significant component of the SAR training also is focused on the protection of privacy and civil rights in the process of handling these reports.[6]

The SAR program is another opportunity to build a standard message that can be exchanged with the pertinent organizations. A specification and a disciplined set of data have been defined in what is called an Information Exchange Package Document (IEPD) that sets out a standard for exchanging information. The intent is that public safety agencies will use this standard for forming messages that can be interpreted by computers that are NIEM aware and thereby automate the information exchange in a timely fashion.

Thoughtful communications CIOs are very aware of the potential value of building standards into the collaboration technologies they need, to lower the cost and risk of implementation. Thoughtful company leaders who develop and implement collaboration technologies have similar objectives. One of the programs that has evolved from this mutual aim is the Public Safety Data Interoperability (PSDI) Program, a joint program of APCO International and the IJIS Institute. Funded by the Bureau of Justice Assistance in the Office of Justice Programs in the U.S. Department of Justice, this program aims to raise the level of awareness of the need for standards dealing particularly with interagency collaboration and information sharing. The PSDI program already has generated a number of reports that serve to define the needs for information exchanges and the best practices that may increase the number of standardized exchanges.[7]

Beyond the better exchange of information that is disciplined by standards, contemporary technology trends provide creative opportunities to support far more effective collaboration. Tactical and strategic information sharing can be enabled by the Internet technology and the same kinds of social networking tools that are now the rave of the day. Professional group dialogue can be done in very large scale by groups that can be created in LinkedIn, Yahoo, and many other existing platforms that people are already connected to. There are also several dozen public domain and supported software packages available to create very effective wikis that can be used for collaborating on a response to an incident or on the acquisition of a new building or just sharing effective practices.

The rapid adoption of video technology for remote surveillance, automatic license plate reading, and sensors are providing new sources of information that also will need to be shared when multiple agencies are engaged in a single event. The techniques and support for such sharing obviously need to be worked out well in advance of their potential deployment, so advanced planning, acquisition, and testing are essential steps toward readiness.

As agencies short on staff begin to explore more effective mutual aid and deeper collaboration, communications centers will be stretched to find more cost-effective ways to share information across jurisdictional boundaries. Technologies such as the new telepresence capability can provide high-definition video with high-definition audio to create meetings across jurisdictions and agencies where the participants can feel like they are working together as if they were in the same room.[8] One of the most dramatic demonstrations of this technology was done by Cisco Systems in 2007, projecting a visual image on stage in an almost holographic experience resembling something out of Star Wars.[9] The more effective way that telepresence stages collaboration can save enormous costs for travel and meeting times but also, more importantly, allows a meeting to be convened in short order instead of waiting for the participants to be rounded up.

We are just beginning a new era in collaboration technology. As those who are responsible for executing public safety missions see the increased need to work together both to be more effective and to reduce costs, the need for better collaboration technology to support these new directions will be enormous. Everything we have talked about in this chapter, in relatively short order, will be viewed as very rudimentary tools that started us on the way, but they mostly will be eclipsed by innovations we have yet to dream about. Our progress will be measured by the extent to which articulate communications CIOs can describe the requirements for technology that supports higher levels of collaboration throughout the public safety world.

PAUL WORMELI is Executive Director Emeritus of the Integrated Justice Information Systems Institute, a non-profit corporation formed to help state and local governments develop ways to share information among the disciplines engaged in homeland security, justice, and public safety. He has had a long career in the field of law enforcement and justice technology. He has been active in the development of software products, has managed system implementation for dozens of agencies throughout the world, and has managed national programs in support of law enforcement and criminal justice agencies. Mr. Wormeli was the first national project director of Project SEARCH, and was subsequently appointed by the President as Deputy Administrator of the Law Enforcement Assistance Administration in the U.S. Department of Justice. He is a member of the Committee on Law and Justice (CLAJ) of the National Academy of Sciences, and in 2010 was named by Government Technology magazine as one of the Top 25 Doers, Dreamers & Drivers in Public Sector Innovation in the U.S.

ENDNOTES

1. Bratton, William and Tumin, Zachary, *Collaborate or Perish!: Reaching across Boundaries in a Networked World*, (New York, Crown Business, 2012)

2. Anonymous, *The Wisdom of the Crowds*, http://en.wikipedia.org/wiki/The_Wisdom_of_Crowds (October, 2011)

3. Anonymous, *Collaborative Software*, http://en.wikipedia.org/wiki/Collaborative_software (January, 2012)

4. APCO International, APCO/CSAA ANS2.101.1.2008, http://www.apco911.org/new/commcenter911/documents/APCO-CSAA-ANS2-101-1web.pdf (August, 2009)

5. McMahon, Kathy, *Next Generation 9-1-1 in 2011*, http://psc.apcointl.org/2011/01/05/next-generation-9-1-1-in-2011/ (January, 2011)

6. Paul, Kshemendra, Nationwide SAR Initiative, http://www.ise.gov/nationwide-sar-initiative , (Fall, 2011)

7. Parker, Scott and Wisely, Steve, *Guide to Information Sharing and Data Interoperability*, http://www.ijis.org/docs/Guide_Info_Sharing_Data%20Interoperability_Local_Comm_Ctrs_FINAL.pdf, (August, 2009)

8. Anonymous, *Telepresence*, http://en.wikipedia.org/wiki/Telepresence, (Various)

9, Indrayam, *Cisco Telepresence Magic*, http://www.youtube.com/watch?v=rcfNC_x0VvE. (November, 2007)

Chapter 6: Public Safety Wireless Communications—Past, Current, and Future

BY CHIEF HARLIN R. MCEWEN

Our police, fire, medical, and other emergency professionals must have access to modern and reliable communications capabilities, including mission-critical voice and high-speed data and video, to communicate with each other and with federal officials across agencies and jurisdictions during emergencies. State-of-the-art public safety communications are essential to provide the public with the protection and security it deserves. Public safety touches every facet of our lives, including the safety of our families and economic growth.

To predict the future, it is often good to reflect on the past. In my lifetime, I have experienced tremendous technological changes that help me look cautiously at the future.

Past

When I started my career as a police officer in 1957, I never imagined the technology we use today. Two-way radio equipment was mostly low band (20-50 MHz), analog, and largely mechanical. We often experienced long-distance interference from radio units thousands of miles away. Base and mobile units were big and power hungry, and had vacuum tubes and few, if any, transistors. Mobile radios were power hungry and put a strain on the 6-volt car batteries of those days, and when you were driving at night, the radio would dim the headlights while transmitting. Radio manufacturers gradually eliminated vacuum tubes and used transistors that allowed radios to become smaller and lighter weight. Car manufacturers switched to 12-volt car batteries with much more efficient generators. When I started, the only portable units were large and heavy with a telephone-type handset, a heavy and very large battery, a very long whip antenna, and a shoulder strap to help manage the weight. Dispatching police and fire units consisted of simple push-to-talk voice conversations.

In that time period, there were no computers. To write our reports, we used manual typewriters that cut holes in the paper, and we used carbon paper to make multiple copies. There was no spell check other than the often-used dictionary; correction tape was used to correct errors. The reports were often quite messy looking. There were no copy machines, and multiple copies of documents were mostly prepared on mimeograph machines, which consisted of typing a master template that cut holes on a film-like sheet, which was then put on an ink drum that printed the copies. The landline telephone system was manual with no rotary or touch-tone dials, and most people shared their phone line with others on a party line. Only one person could use the phone, and it was not uncommon for people to eavesdrop on each other's conversations.

In the years that have passed, I have seen many improvements in public safety communications, including the development of integrated circuits and other electronics that allowed production of hand-held portable radios, the movement from analog to digital systems, and development of computerized or software-defined radios. For today's public safety personnel, these products are commonplace and taken for granted.

I have served as Chairman of the Communications & Technology Committee of the International Association of Chiefs of Police (IACP) for the past 35 years and during those years have been a part of the many discussions about improving public safety communications. During that time, we advocated for additional public safety radio spectrum to meet the increasing needs of our public safety personnel.

On June 25, 1995, the Federal Communications Commission (FCC) and the National Telecommunications and Information Administration (NTIA) established the Public Safety Wireless Advisory Committee (PSWAC). The task for PSWAC was to evaluate the wireless communications needs of local, tribal, state, and federal public safety agencies through the year 2010; identify problems; and recommend possible solutions. I was fortunate to serve as a member of the PSWAC Steering Committee. The Final Report of the PSWAC was issued on September 11, 1996, (exactly five years before the tragic events of 9/11/2001) and called for the allocation of 97.5 MHz of additional spectrum for public safety use. As a result of the PSWAC Report, in 1997 Congress directed the FCC to allocate 24 MHz of spectrum for public safety in the 700 MHz band. This spectrum was designated to be abandoned by television broadcasters who would be converting from analog to digital signals. Each analog TV channel was 6 MHz in width, and TV channels 63, 64, 68, and 69 were allocated by the FCC for public safety use.

In 1999, the FCC established the Public Safety Coordination Committee (NCC) to advise the Commission on a variety of issues relating to the use of the 24 MHz of spectrum allocated to public safety. I also was fortunate to serve as a member of the NCC Steering Committee. The first meeting was held on April 6, 1999, and the last meeting was held on July 17, 2003. In its final report, the NCC recommended that half of the 24 MHz of spectrum be designated for public safety narrowband voice channels and half be designated for wideband data channels.

On July 31, 2007, the FCC issued an order that changed the channel assignments within the 24 MHz of public safety spectrum in the 700 MHz band. The 12 MHz of narrowband voice channels that were previously broken into four 3 MHz groups were consolidated into two 6 MHz groups and moved within the public safety allocation. The 12 MHz of wideband spectrum that was previously broken into two 4 MHz Groups was changed to two 5 MHz broadband groups.

As a result of that order, the FCC issued a nationwide public safety broadband license for the 10 MHz of broadband spectrum to a newly formed not-for-profit corporation called the Public Safety Spectrum Trust (PSST). The PSST consists of the representatives of 15 national public safety organizations.

Current

I will not go into detail on the lengthy activity that took place between 2007 and 2012, but will simply say that after almost five years of ongoing discussions, Congress, on February 17, 2012, passed a bill titled the "Middle Class Tax Relief and Job Creation Act of 2012," and within that bill was a section titled "Public Safety Communications and Electromagnetics Spectrum Auctions." President Obama signed the bill on February 22, 2012, setting in motion a number of legislated requirements for the development of a new Nationwide Public Safety Broadband Network (NPSBN).

The legislation sets the foundation for the next generation of public safety wireless communications based on Long Term Evolution (LTE) broadband technology that is already being implemented for public use by several commercial carriers. It provides for an additional 20 MHz of broadband spectrum (upper 700 MHz band D Block) allocated for public safety to provide a total of 20 MHz of broadband spectrum necessary to make efficient use of LTE technology. It also provides $7 billion in federal funding that will be realized from future incentive-based auctions of spectrum and not from tax-based funds. The legislation establishes new governance for the NPSBN called the First Responder Network Authority (FirstNet), which will have wide discretion on the planning and implementation of the NPSBN.

The expectations are that the FirstNet Board will develop a plan to provide additional funding that will come from public/private partnerships. The NPSBN is expected to have greater reliability, security, and coverage than commercial carriers currently provide while at the same time giving public safety access to the latest commercial technologies. Public safety will manage priority access within its own network without competing for spectrum resources on the public networks that are often crowded and unavailable during major events and emergencies. And it is expected that eventually there will be a satellite component of the NPSBN that will provide coverage when terrestrial service is disrupted or in areas where terrestrial service never will be available.

There are issues that are of concern to the public safety community related to the lack of understanding of the capabilities of commercial broadband technology like LTE in regard to public safety mission-critical voice. Andrew M. Seybold is a well-respected communications industry analyst and a longtime public safety advocate who is known for his ability to forecast trends. I personally have high regard for Andy. He has been one who has sounded the alarm and written extensively about the misinformation swirling around us concerning the mission-critical voice capabilities of network-centric commercial broadband technologies.

Andy has written several articles relating to public safety mission-critical voice and has done a good job explaining the difference between cell phones that use full duplex mode (transmit data in two directions simultaneously so both parties can talk at once) and typical land mobile two-way radios that use simplex mode (push to talk, allowing only one person to talk at a time). Simplex mode can be made through a repeater system, called simulcasting, that is a network-centric approach or direct mode (one unit to another unit or one unit to many units) without a network. This is commonly referred to as talk-around or tactical mode and is often the lifeline of public safety responders when they cannot or do not want to access a network.

The issue here is that many citizens and many local, state, and federal appointed and elected officials have the false idea that the current public safety radios can be replaced by cell phone-type devices or broadband devices. That is simply not true. Current mission-critical public safety voice systems are the lifeline of public safety responders and cannot be abandoned when there is no path forward that promises they can be replaced by emerging network-centric broadband technologies.

Public safety must have wireless devices that always work and allow for one-to-one and one-to-many communications, so when they need to call for help or perform their critical duties, they can always communicate. Network-centric communications do not always work because the emergency responder's radio device cannot always reach a network, or the network may be overloaded. I realize that some say that there is no radio that will always work, but I am talking about providing devices that will work without a network and normally can always reach directly another nearby unit.

To close the discussion of the current situation, public safety faces many challenges addressing the legislative requirements. The development of a plan for the Nationwide Public Safety Broadband Network and the subsequent implementation of the plan will not be an easy task. This is a complex matter that will require the inclusion and participation of many people at the local, tribal, state, and federal levels, and many people from industry if this network is to be successful.

Future

The future is exciting. The next 10 to 15 years will see the development and implementation of the NPSBN, and this development will bring exciting new data services and improved interoperability.

Public safety organizations like the International Association of Chiefs of Police (IACP), National Sheriffs' Association (NSA), International Association of Fire Chiefs (IAFC), National Association of State EMS Officials (NASEMSO), and the Association of Public-Safety Communications Officials-International (APCO) continue to provide focus on public safety communications issues and represent the ongoing needs of public safety personnel.

The Office of Emergency Communications (OEC) within the U.S. Department of Homeland Security (DHS) has been extremely helpful in developing regional and state coordination and educational programs to improve public safety communications and interoperability. The OEC relies strongly on the input and advice of its advisory process, called the SAFECOM Program. The SAFECOM Program is composed of local, state, and federal public safety professionals, as well as representatives of local and state government organizations that form an Executive Committee (EC). The SAFECOM EC is supported by a larger group of public safety and government representatives known as the Emergency Response Council (ERC).

The Office of Interoperability and Compatibility (OIC) within the DHS Science and Technology Directorate is the research program supporting improvements in public safety communications, and it also has been extremely helpful as new technology emerges and offers promise to new and exciting opportunities for public safety.

The Public Safety Communications Research (PSCR) program located in Boulder, Colorado, has been critical and will continue to be critical to providing technical support, testing, and research of emerging public safety communications equipment and systems. The PSCR is a joint program of the National Institute of Standards and Technology (NIST) and the National Information and Telecommunications Administration (NTIA) within the U. S. Department of Commerce.

The National Public Safety Telecommunications Council (NPSTC), a federation of 15 public safety national organizations, created the Public Safety Assessment of Future Spectrum and Technology (AFST) Working Group in August 2009 to identify public safety communications requirements for the next 10 years. It is a follow-on to the 1996 PSWAC report. The final report titled "Public Safety Assessment of Future Spectrum and Technology" (PSAFST) will be released in 2012 and will cover the period 2012 through 2022. The report is well done, and a draft was circulated for comment from the public safety community and communications industry prior to finalization. The AFST Working Group has done an excellent job in conducting research and national questionnaires to collect information from public safety personnel on the real current and future communication needs of public safety.

The PSAFST report reviews the 1996 recommendations in the PSWAC report and comments on whether those recommendations have been implemented, and it reflects on those recommendations in today's environment. The report is broken into three key areas: operations, technology, and spectrum. It sets forth more than 20 key findings and recommendations that clearly identify the significant investments that have been made over the past 10 years to improve public safety communication and interoperability, and it also identifies the critical needs and challenges of the future.

In summary, reliable public safety communications are critical to the ability of public safety personnel to provide their services to the public in a safe and efficient manner. Reflections on the past and current situation and preparing for the future are important as we move forward into the next generation of public safety communications technologies.

HARLIN R. MCEWEN serves as the Chairman of the Communications and Technology Committee of the International Association of Chiefs of Police (IACP), a position he has held for the past 35 years. He started as a patrol officer, and after progressing through the ranks, served as a Chief of Police for more than 20 years, last serving as Chief in the City of Ithaca, N.Y. Following Chief McEwen's retirement as Ithaca Police Chief, he served as a Deputy Assistant Director of the Federal Bureau of Investigation with his office located at FBI Headquarters in Washington, D.C. He is a Life Member and Honorary President of the IACP, Life Member of the National Sheriffs' Association (NSA) and Life Member of the Association of Public-Safety Communications Officials-International (APCO).

Chapter 7: Concepts on Information Sharing and Interoperability

BY JOHN M. CONTESTABILE

This document was prepared under an Urban Area Security Initiative (UASI) grant to the National Capital Region (NCR) from FEMA's Grant Programs Directorate, U.S. Department of Homeland Security. Points of view or opinions expressed in this document are those of the authors and do not necessarily represent the official position or policies of FEMA's Grant Programs Directorate or the U.S. Department of Homeland Security. Portions of this paper are Copyright IEEE and have been presented at the IEEE Conference on Homeland Security Technology 2011 as "Information Sharing for Situational Understanding and Command Coordination in Emergency Management and Disaster Response," Robert I. Desourdis, Jr. and John M. Contestabile. Used with permission from IEEE.

This paper addresses a conceptual framework for sharing information across jurisdictions, agencies, and public safety disciplines. It was developed as part of the NCR jurisdictions' (Maryland, Virginia, and the District of Columbia) interoperable communications programs. The paper explores why information sharing is important to successfully dealing with large-scale events and how a lack of public safety communications systems interoperability is a major impediment. It describes how a conceptual framework of information layers (the Data, Integration, and Presentation layers) is useful to developing solutions to the lack of interoperability. It further describes a concept of operations whereby Integration layer applications can form the core of a "Common Operating Picture," which can provide information to field personnel at the scene of an incident as well as the public. Some regions of the country have implemented tools consistent with this concept (notably the National Capital Region), while elements of this concept can be found in others. An inducement for jurisdictions to participate in such an information-sharing framework is that they can gain access to a wide array of information to which they would otherwise not be entitled, and they can reduce the overall cost of such systems by sharing the infrastructure and system expenses across the regional partners. Additionally, it is recognized that governance and security issues become increasingly important in such an information sharing environment.

Introduction

There are hundreds of thousands of incidents that occur every day in the United States, from simple/frequent incident events like automobile accidents, train derailments, theft, weather incidents, to catastrophic/infrequent incident events like the 9/11 terrorist attacks, Hurricane Katrina, the Minnesota I-35W bridge collapse, and the December 2004 tsunami, to name just a few. The number of participants and resources required to respond and recover, and the complexity of their roles and responsibilities, are significantly greater and more difficult for a catastrophic incident than for a simple incident. Understanding the information needs between these different scale incidents will provide some insight into how various agencies and jurisdictions can better design their information systems. In short, how these systems are designed will directly correlate to the ability to share information across agencies, jurisdictions, and disciplines. That is, the design determines the systems' level of interoperability. This paper will discuss the all-hazards operational incident response and the implications for information sharing as well as propose a conceptual framework to improve interoperability based upon three layers: data, integration, and presentation.

Incident scale and its implication for information sharing

While experts can identify roles for the nation's first responder community, it is important to note that these roles are not always fulfilled on each and every incident that occurs. From an "all-hazards" perspective, incidents vary widely, from a relatively minor "fender bender" on the Interstate Highway System all the way to a terrorist event on the order of magnitude of 9/11, or a natural disaster such as Hurricane Katrina, or the 2004 Indian Ocean tsunami.

This incident scale schema (Figure A) characterizes the scope of the response to an incident as local, regional, state, or national. This somewhat simplistic characterization will have a bearing on the number and type of agencies responding. It is within this context that a discussion of information sharing must occur because providing relevant information to the right people in a timely manner will determine the ability to deal with the event successfully.

Incident scale is directly associated with the level of public preparedness for a given type of incident as well as the complexity of the response coordination. For example, for the fender bender-type traffic incident, the number of responding agencies involved is quite low; often only a police cruiser and officer will respond to the scene. In this example, public preparedness is high, as the type of incident is fairly commonplace, and the complexity and need for any other agency's involvement is low. Residents typically learn of this event through radio traffic reports, and a common reaction may be to "get off the highway an exit or two early," avoid the inevitable traffic tie-ups and "go home the back way." As for the first responder, the police officer would call in the license plate number to dispatch, talk with the motorist(s) involved and, barring any significant injuries, call a tow truck and perhaps stay on the scene until the truck arrives. The scene would be cleared from such an incident in less than two hours, and the disruption to traffic and the surrounding communities would be minimal. This type of incident is shown as a local incident in Figure A. The few agencies involved, the minimal impact to the public, the lack of "ripple effects," and the relatively short clearance time make it a localized incident.

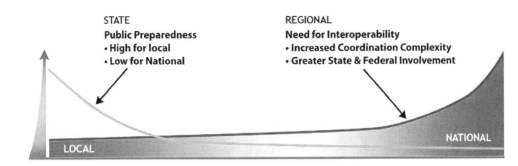

Figure A. Incident Scale/Public Preparedness

To continue with this example, should the license plate check or the check of the driver's identification surface the fact that the vehicle was stolen or that the driver was found on a watch list of some kind—i.e., outstanding warrant or on some sort of "person of interest" file—the response scenario would be much different. It is likely that backup forces would be called and, depending on the seriousness of the information uncovered, a police helicopter might be deployed overhead. It also is not uncommon for a lane or more of traffic to be closed while the vehicle and its occupants are examined. If this were the scale of the response, it would likely attract media attention, and senior leaders in all the organizations involved would make inquiries and need to be briefed. This would necessitate additional communications efforts from the scene to "headquarters," and agency public information officers (PIOs) would become involved.

This example illustrates the fact that incidents rapidly can become something more significant than an initial assessment may indicate. If this scenario were to occur during "rush hour" and/or the incident lasted more than two hours, it is likely that this could be classified as a regional event as the ripple effects on the transportation system (i.e., the resulting backup from lane closures during rush hour, those individuals that take a different route home, and those that elect to take another mode of transportation as a result) would extend far beyond the incident scene. The time to clear the scene would be extended, the media could be covering the event "live," and system owners/operators as well as the response entities would have to provide updates and briefings. As more agencies respond to the scene, some form of incident command would have to be established. The incident scale, as this event escalates, grows.

Some incidents can be classified as statewide events almost from the onset. For example, the threat of a hurricane, given the usual wide swath of impact, likely would be considered a statewide incident. Statewide events in this graphical schema usually would involve the activation of the state Emergency Operations Center (EOC). In these types of scenarios, multiple agencies are involved, incident command must be established, communications interoperability is much more important, and the need for a coordinated response across various agencies or disciplines (police, fire, EMS, transportation, etc.) and jurisdictions (town, city, county, state, and federal) is paramount. The success or failure of the response to a statewide event is in large measure determined by how well this coordinated response unfolds in a timely fashion.

Some events can be classified as national events. The events of 9/11 clearly were national in scope, as the air travel network was shut down for a period of time, and the whole country felt the impact of the crisis. The impact of that event extended beyond the transportation system to the financial markets. Hurricane Katrina is also categorized as a national event, because it affected interstate commerce, and many states across the country absorbed refugee populations from the states more directly impacted. The supply chain interruptions extended far beyond those states immediately impacted for months afterward. It is in these types of national events that the federal response is most prevalent and most necessary. Events of this order of magnitude evoke the Stafford Act and a Federal Emergency Management Agency (FEMA) response. Should the event have a terrorist connection, the Federal Bureau of Investigation (FBI) as well as elements of the Department of Homeland Security (DHS) would be involved. The main point in this type of incident is that multiple agencies from the federal level to the state and local level would be involved in the response, and the communication needs become exponentially more complicated.

It also is important to note that once an event is seen as a national event, it does not eliminate or reduce the role of local, regional, and state assets. As the saying goes, "All incidents are local." That is, the local first responders will be involved at the outset and will remain involved over the life of the incident. However, additional assets will become engaged from other jurisdictions and disciplines.

This incident scale schema helps to frame the different types of incidents that responders and the emergency management community will face and the resultant complexities that will emerge. It illustrates that larger scale events will have communications, organizational, resource, and coordination challenges that make effectively dealing with such events problematic. While roles and responsibilities can be defined in advance, they may not be fulfilled unless the incident warrants the involvement of a particular agency or entity. And, general roles and responsibilities must be tailored to the particular event. All of this has a bearing on the need for certain types of information, who should receive it and when, how the information is transmitted and displayed, etc.

A final complication that overlies this schema is the issue of time. As mentioned earlier, incidents can escalate rapidly and become something much more complicated than first thought. For example, returning to the "fender bender" local incident, what if the vehicles involved were a passenger car and a tanker truck carrying hazardous materials? And, what if the tanker truck was damaged in such a way as to begin leaking product that created a plume, threatening a nearby school? The challenges to respond promptly, size up the situation, establish communications, establish a command structure, obtain weather information, warn and evacuate (or shelter in place) the school and neighborhoods involved are enormous. In a moment, a single variable in an otherwise common local incident can make time the critical factor on which lives depend.

The thinking of incidents as local, regional, statewide, or national helps responders and other involved agencies/jurisdictions grasp the inherent complexities as one moves from left to right on the graphic as an incident escalates. Those agencies and jurisdictions require established communications and command and control systems that can adapt as quickly as the event itself may escalate. Understanding roles and responsibilities in this context will help those involved to recognize the limitations and challenges of current systems and identify gaps where improved protocols, communications systems, and resources are needed. Successfully dealing with an emergency incident involves getting the right information to the right people at the right time.

Public safety communications interoperability

A significant barrier to getting information to those that need it in a timely manner has been referred to as a lack of communications "interoperability." (See: http://www.safecomprogram.gov/SAFECOM/interoperability/default.htm.) That is, systems cannot share information readily with other systems. These systems could either be voice communication systems (such as an 800 MHz system user that cannot talk with a 450 MHz system user because of the different frequency bands) or data systems (such as Geographic Information Systems—GIS, Computer Aided Dispatch Systems—CAD, or Traffic Incident Management Systems—TIMS) that utilize different data formats, programming code, or lack standards for information sharing. This lack of interoperability among systems impedes the flow of information across jurisdictions (e.g., from a county EOC to a state EOC), agencies (e.g., from the Department of Motor Vehicles to the local police field units), and disciplines (e.g., between police and EMS).

So, if the incident scale discussion illustrates that sharing information across jurisdictions/agencies/disciplines is key to successfully dealing with larger scale events, and that information sharing is impeded by a lack of interoperability of the communications systems involved, then reducing the causes of interoperability should improve information sharing. However, reducing interoperability problems is much easier said than done, for numerous reasons. There has been much effort put into this problem over the past several years, including the naming of Statewide Interoperability Coordinators (SWICs), the development of State Communications Interoperability Plans (SCIPs), targeted grant programs (Interoperable Emergency Communications Grant Program—IECGP, for example), as well as the publication of considerable federal guidance. (See: http://www.safecomprogram.gov/SAFECOM/.)

One of the reasons solving the interoperability problem has proven so difficult is that it is not solely a technical problem. Agencies have not purposely built systems that would not work with other systems, but rather they built systems to meet their particular business needs within normal budget limitations. If it was not determined to be a critical need to share information with another agency, then scarce dollars were not allocated to providing that connection. And while that may be true for a local event (as discussed above), that does not hold true if the event scale could be considered a regional, state, or national event. In those cases, sharing information widely is key to successfully responding to and recovering from that event. So, a lack of interoperability remains an issue in existing systems (and even in planned systems) because of a lack of perceived need to share information widely (because an agency may only participate in more than a localized event only a few times a year) or because of insufficient funding to adjust the project to make the system more interoperable.

In addition to a perceived lack of need or lack of funds to build a more interoperable system, there are several other factors that need to be considered. The Department of Homeland Security "Safecom" Program has identified five factors that have a bearing on interoperability: governance, standard operating procedures, technology, training and exercising, and usage. (See: http://www.safecomprogram.gov/SAFECOM/tools/continuum/default.htm.) The "Interoperability Continuum" illustrates that there are degrees of interoperability and that some progress across all these factors must be made in order to improve interoperability.

While the continuum is quite useful in understanding the impediments to interoperability, it is not detailed or specific enough to provide a framework for achieving interoperability. While very specific details must be left to the locale in which interoperability is being addressed (that is, the governance, standard operating procedures, technology, etc., in that area), a technical framework for achieving interoperability can be articulated. The remainder of this article hopes to provide a framework for achieving data interoperability.

How might communications interoperability be achieved during emergency events?

As mentioned above, solving the interoperability problem is not just a technical issue. All too often, money has been spent on a technical solution (a "black box" solution) only to find that the solution does not meet the need of end users. This matter is more than the technologists doing a better job of requirements gathering. Solving the interoperability challenge involves navigating human relationships and issues of trust, and it must be approached in that fashion. The importance of trust has been raised in many forums (see the All Hazards Consortium at www.ahcuas.org for example), and the lack of trust will impede information sharing.

Recognizing these challenges, some success can be had if the problem is approached sequentially from a people, process, and technology standpoint. That is, the people from different jurisdictions/agencies/disciplines must come together and work through a process whereby they can understand each other's need for information, and trust can be developed between the parties. Only then can a technology approach/solution be identified and applied. Often times, grant deadlines, consultant schedule constraints, preconceived notions as to the "right" solution and a general lack of understanding of this dynamic work against giving the people and process steps sufficient time to develop a creative and workable technical solution.

While solving the interoperability challenge is not solely a technical matter, technology still is an important part of the solution. In fact, there is a dynamic between technical and non-technical factors that is somewhat symbiotic. It is all too easy for the participants in the process to pay lip service to sharing information if they know that there is no technical way for them to do so. Once a technical approach has been identified (if not actually applied), the participants must own up to the commitment to share data by investing and working toward the solution. This is the turning point in the process when the participants have the "ah ha" moment and identify an approach, architecture, or solution that everyone can buy into, which creates conviction and momentum. Only then will the project have the potential for success.

To summarize, successfully dealing with larger-scale events requires sharing information widely, and a lack of interoperability between the systems that hold that information is a major impediment to success. There are factors beyond the technology that have a bearing on solving the interoperability challenge, and the people who have the need to share information must work through a process of discovery to identify an appropriate solution that works in their setting. Experience in developing solutions in this space suggests that there is a pragmatic approach to this problem that is applicable in most settings. The proposed conceptual framework that follows would provide for improved information sharing that could link various operation centers as well as field units at the scene of an incident.

A conceptual framework for information sharing and improved interoperability

Consider a conceptual interoperability framework in which there are three levels that can be applied to most settings where interoperability is desired and can be achieved with minimal impact to existing systems. The three layers comprising the framework include:

- the data layer,
- the integration layer, and
- the presentation layer.

[Diagram showing three stacked layers: PRESENTATION LAYER (top), INTEGRATION LAYER (middle), and DATA LAYER (bottom).]

Figure B.

The data layer

At the bottom of Figure B lies the data layer where all the various data sets and applications spread across various jurisdictions/agencies/disciplines reside. Local data sets (for example, property patterns, zoning, locations of fire hydrants, school building plans, crime statistics, water supply and storm water systems, etc.), regional data sets (such as traffic network volumes, landfill information, wastewater treatment systems, etc.), state data sets (such as health records, social services, state roadway data, environmental information, etc.), as well as federal data sets (such as geospatial, aerial imagery, and crime statistics—for a more comprehensive list of examples, see www.data.gov) are contained in this layer. While the location of this data can vary from place to place (that is, which agency or jurisdiction is responsible), there is no doubt this data exists in every location and that some agency is responsible for creating it, tracking it, and maintaining it for some legitimate business purpose.

Typically, these systems lie behind agency firewalls, were built with some level of customized code (even if off the shelf software/applications were used), and are designed for agency use, not designed to share information with others outside the agency or beyond the firewall. In fact, Chief Information Officers (CIOs) of these agencies often are unwilling to share information from these systems to others outside the firewall because of costs and legitimate security concerns. Additionally, in the case of public agency data systems, these systems are often older, large, complex systems (think of driver's license, health care, and voter systems, for example) in which CIOs are wary of creating interfaces to other agencies for fear of the effect it will have on the stability of the rest of the system.

One method of improving data sharing would be to create interfaces between all the disparate systems at the data layer but, for some of the reasons noted above, this is problematic. Additionally, if one were to provide for interoperability at this layer, it would result in a multitude of "one off" connections. For example, if county police agency A wished to share information with an adjoining county B, they could build a custom interface between their systems. If county police agency C also wanted to see that information, an interface would have to be built with that agency, and so on. Ultimately, there would be multiple different interfaces between each of the agencies that wanted to share information. One can appreciate how a CIO would not embrace this approach by having to develop, fund, and support multiple interfaces to the same system.

The integration layer

A more artful approach, in keeping with today's networked architecture, would be for those data layer systems to publish once to an integration layer tool. Those agencies/jurisdictions and disciplines who need to see that data, could now look to the integration layer tool to see that information linked to other agencies with like data. To return to the previous example, police agencies A, B, and C would all publish their data once to an integration layer tool so that if any of the agencies desired to see any of the other agency's data, they would look to the integration layer tool; not to the other agency. Done properly, this would be transparent to the individual agency; that is, each agency still would use their native system, but the results would be published to the integration layer tool out of the "back end" of the system.

Of course, publishing data from the data layer to the integration layer would need to respect network protocols, security requirements, and the appropriate standards for that data. The concept would be to publish the data out of the typically proprietary, customized, legacy/mainframe environment from which it came in the data layer into a web-enabled, Internet Protocol (IP) and standards-based open environment in the integration layer.

With data having been published into the integration layer, interoperability then can be achieved by connecting the various tools found in that layer. Because these tools are more amenable to integration, they can be connected, and data can be shared across these tools so that it can be seen in a larger context. Unlike trying to achieve interoperability at the data layer, providing only a few interfaces between a handful of key integration tools is feasible. The presentation of this three-layer schema can be seen in the Figure C on the following page.

What tools should be provided in the integration layer?

This question is akin to asking what data is needed during an emergency. While one cannot give a complete answer due to the unique information sharing needs of each incident,

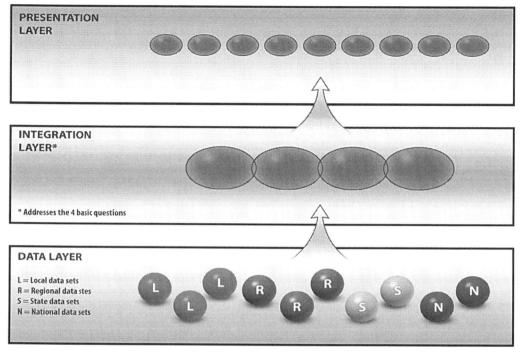

Figure C.

there are certain information needs which are almost always required. Typically, four questions need to be answered:

- Where is it?

- Can we talk about it?

- What do we know about it?

- Can we see it?

These questions have ramifications for four types of data or capabilities:

- Geographic Information Systems (GIS)

- Voice Communication Systems (as well as Critical Incident Management Systems – CIMS)

- Access to disparate data sets (such as sensors)

- Video systems

Thus, the integration layer tools must address (at least) the four types of desired data: GIS, CIMS, Sensors, and Video—as well as other data sets. Some sort of application or tool that can "ingest" information of that type and aggregate it with other like information as well as share it horizontally with the other tools in the Integration layer is what is needed. Additionally, now that the disparate data has been aggregated and integrated, it may be necessary to overlay analysis and decision support tools to make better sense of this wide ranging set of data. Revisiting the Conceptual Interoperability Schema graphic, the integration layer tools could be labeled as noted below.

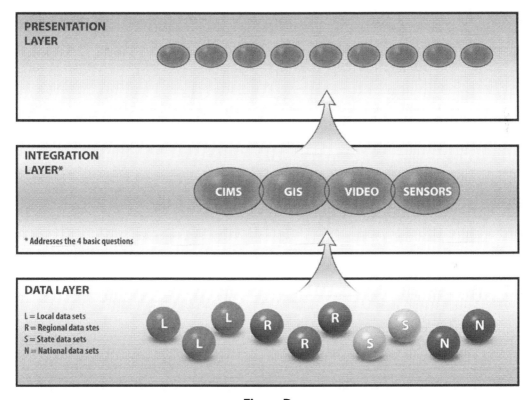

Figure D.

The presentation layer

Now that the data has been published into a handful of integration tools and those tools have been connected to achieve interoperability, the fused data needs to be "served up" to allow visibility across agencies/jurisdictions and disciplines by publishing into the presentation layer using a variety of channels—from telephony, to web-based, to Short Message Service (SMS), and Multimedia Messaging Service (MMS)—both wired and wireless. This will allow the information to be delivered to those that need it (via push and pull methods) across emergency operation centers, incident command posts, responders, as well as the public. The presentation layer can be used to distribute the information beyond the data owners that have provided it to the integration layer and can take advantage of existing social networking tools to extend their reach.

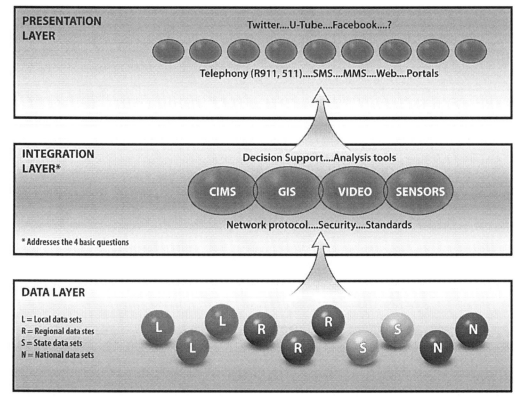

Figure E.

Operational model

With such an information-sharing schema in place, the participating partners/agencies that provide information to the integration layer can then see their information in relationship to the other partners. For example, the GIS tool would show the location of incidents listed in the CIMS software log, as well as links to the video cameras and other sensors in the vicinity as well as across the region. The GIS tool also would have multiple layers of the information available on roadways, schools, shelters, evacuation routes, transit/rail systems, parks, utilities, critical infrastructure, etc. This tool could form the basis of a Common Operating Picture (COP), which all the partners could see—with information updated and published in near real time.

Given that this suite of integration tools would have the most up-to-date information during an incident, this COP could be the information-sharing engine that bridges operations centers and field units at the scene of an incident. Today, there are many variety of operations centers—such as state/county/municipal Emergency Operations Centers (EOCs), Traffic Management Centers (TMCs), fusion centers, as well as utility companies and transit Operations Control Centers (OCCs)—functioning on a 24x7x365 basis. However, there is typically no common software platform to which they all can look to have a shared understanding before/during/after an incident. The integration layer tools would provide a COP and, as such, a vehicle for collaboration across centers and a method to respond to requests for information from the field units. This information-sharing framework would be a way to engage these various centers in supporting the field personnel and the incident command system while providing a much needed collaboration tool and COP. Such a conceptual operational model can be seen below.

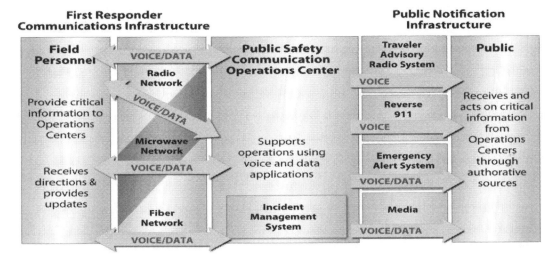

Figure F. Operational model.

While the discussion above relates the possible uses/benefits to the various operations centers and field personnel, this suite of tools in the integration layer also would provide benefit to communicating with the public. Selected information from this suite could be published to the presentation layer and distributed via:

- the web and/or various social networking tools

- telephonically through reverse 911 or 511 or SMS

- Traveler Advisory Radio (TAR), or

- the proposed Integrated Public Alert and Warning System (IPAWS)

These varied methods of information distribution via automated means would speed the dissemination of authoritative information to the public during incidents when timely and accurate information sharing is critical.

Efforts to implement the conceptual framework

Some jurisdictions are building systems/solutions that comport to this Conceptual Interoperability Schema. For example, the National Capital Region (NCR, which includes the District of Columbia, Northern Virginia and a portion of Maryland), using Urban Area Security Initiative (UASI) funds, has put in place many tools in the Integration layer to achieve information sharing. They also have invested in developing a regional fiber network (called the NCRnet) and protocols for information sharing (called the Data Exchange Hub – DEH). Information on NCRnet and DEH can be found at www.ncrnet.us.

An original graphic (developed by others in the mid-2000s) depicting these applications organized within the conceptual framework can be seen below. Regional tools such as LINX (Law enforcement Information Network Exchange), WebEOC, RITIS (Regional Integrated Transportation Information System), CAPWIN (Capital Wireless Information Network), HC Standard and a regional GIS tool are all integration layer applications that aggregate like information from the data layer for a variety of end users. These applications are connected in some cases and plan to use the NCRnet and DEH for transport so as not to rely on an Internet connection during emergencies. Some of the presentation layer tools are in place, such as Roam Secure/RICCS (see http://riccs.mwcog.org/faq.php), but much of this layer of the framework is still being built out.

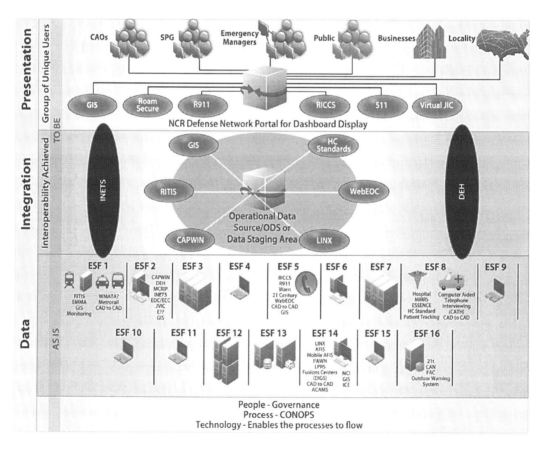

Figure G. National Capital Region Information Sharing Framework.

What are the benefits and challenges?

The benefits of developing information-sharing systems according to this conceptual framework are that participating agencies will have access to a wealth of information in the integration layer upon which to make better decisions before, during, and after an emergency incident. Incident commanders are routinely challenged in most every emergency incident they face to make decisions in the absence of information and, while it would be naive to think a commander would ever have all the information needed, such a schema would improve considerably the information at his/her disposal. Additionally, this information may help save lives of both responders and victims, and save time (to formulate decisions and take actions). Additionally, this approach allows agencies to continue to use their legacy systems in the data layer, while taking advantage of other tools/applications in the integration layer for improved situational awareness.

The challenges are that it requires agencies/jurisdictions and disciplines to agree to share their data as well as fund and share the needed integration layer tools. As noted above, this requires that those involved see the need to share information and develop a certain level of trust that they can do so in a secure fashion. Regarding funding, another challenge is that integration layer tools are shared, and yet we continue to budget funding by agency and jurisdiction. Regional grant funding has been able to bridge that gap, yet ongoing funding for sustainment can be a challenge without commitment from the participants.

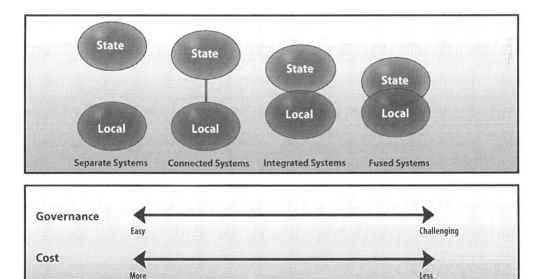

Figure H. Benefits and challenges of integrated systems continuum.

The incentives for information sharing are that costs can be driven down by sharing infrastructure/systems and pooling resources. The graphic below illustrates that interoperability improves as systems evolve from separated to connected to integrated and ultimately fused systems. But, as systems become more connected, governance becomes increasingly important. With shared systems, an individual agency's ability to make changes to that system is constrained, and a mechanism to adjudicate disputes must be in place. There also may be a need for memoranda of understanding (MOU) or agreement (MOA) to set up the necessary governance and ongoing funding.

While the benefits of this information sharing framework are potentially quite considerable in information availability and cost savings, the obstacles of governance, agreements, and long-term funding are likewise formidable.

Summary

- The ability and speed with which you can share information across agencies/jurisdictions and disciplines during an emergency will determine how well that incident will be managed. This is why communications interoperability is important; as it is the key impediment to sharing information across the various incident stakeholders.

- Achieving interoperability at the data layer, by connecting systems/data sets at the individual agency level is not prudent, scalable, manageable, or realistic. Thus, creating an integration layer with a handful of key applications/appliances that can consume published data (in near real time, optimally) is the strata at which interoperability can be achieved.

- Of course, publishing into the integration layer must respect network requirements, appropriate standards for the data being published, and security. The data owner must be able to set the security level of their data and, thus, which users can view that data.

- Once the data is consumed into the integration layer, it is shared across the other applications so as to achieve interoperability and contribute to a more complete operating picture during an incident. Analytical and decision support tools also are useful in this layer to bring key information to the decision makers' attention.

- Benefits of creating an integration layer are: access to data across agencies/jurisdictions and disciplines, improved interoperability, and potentially reduced cost overall. But, it is recognized that governance becomes more important in shared systems.

- Applying these concepts to any particular region will result in some variation, but the National Capital Region has built many of its existing systems consistent with this model.

- The presentation layer is important to distributing the integrated data to end users and leveraging private resources, such as social networking tools. While examples can be found of the data and integration layer concepts, development of the presentation layer remains largely underdeveloped at present.

JOHN M. CONTESTABILE is the Assistant Program Manager for Homeland Security for the Johns Hopkins University/Applied Physics Lab. He joined the Lab in July 2009 after retiring from the State of Maryland Department of Transportation (MDOT), where he was Acting Assistant Secretary for Administration as well as the former Director of the Office of Engineering & Emergency Services. In that capacity, he had been responsible for emergency planning/response/recovery activities for the MDOT since 1996. In addition to working at MDOT, Mr. Contestabile was named by Governor O'Malley as the Director of the Maryland State Communications Interoperability Program, reporting to the Superintendent of the Maryland State Police. Previously, Mr. Contestabile served on assignment with the Governor's Office during the Ehrlich administration as the Acting Deputy Director of the Office of Homeland Security. He also belongs to several organizations, including the All Hazards Consortium, the American Society of Civil Engineers and the Baltimore Chapter of the Women's Transportation Seminar (WTS). Mr. Contestabile received his Bachelor of Science Degree in Engineering from Worcester Polytechnic Institute in Massachusetts and holds a Master of Business Administration Degree from the University of Baltimore.

Chapter 8: If You See Something, Send Something— Public Safety and the New Media

BY DR. ALAN R. SHARK

Just as the commentator was encouraging people to send in their eye-witness accounts of the dramatic bus collision, a police chief from a small community in the Midwest was complaining at a public safety technology seminar that he wished "those people with all those fancy smart phones would stop sending us stuff, and simply call 911!" He went on to add that they had no capacity to view all the pictures and videos, and he was afraid about any legal liability in not being able to handle social media. Another official in the same audience stated a similar concern and wanted to wait for the next generation of 911 (NG 911) to handle such issues. He went on to say that the technology is not there, and he lacked the budget for new equipment, staff, and training.

Meanwhile, the commentator actively seeking for more pictures and eyewitness accounts was of course a TV news anchor, and the appeal was far from that of any local government official. To be fair, many localities have indeed embraced social media—mostly in the form of broadcasting information through such mediums as Twitter and Facebook as well as other "push services." Social media platforms have proven to be effective for outbound communications and alerts. Yet, social media by its very name "social" infers much more then one-way communication. Social media is about engagement, interactive dialogues, developing and maintaining relationships. It's also about creating and sharing content—not simply reading.

Social media, by and large, has been embraced by a growing number of state and local governments, and its purpose is geared to what some are now calling "civic engagement." We are seeing the start of an emerging civic engagement ecosystem in cyberspace. Used effectively, Web 2.0 and civic/social media encourage residents to take a more active role in local and state government. They provide a platform to streamline government communications and services through knowledge bases, access to networked resources, real-time data communications and reporting, wireless access, multi-media delivery, integrated self-service options, location-based services and cross-agency information sharing. Nevertheless, the public safety community has been wary and, in fact, has encountered some rather interesting roadblocks—often of their own making.

In general, public managers are increasingly looking for new ways to engage the public as a means of improving communications and restoring trust. We know, for example, that at the end of 2010, there were 302,947,098 mobile phones, or 97.4% penetration. The research firm Nielson predicted that 2011 would be the year when smartphone penetration would surpass regular mobile phones. However, that prediction was offered before the unexpected and explosive growth of the iPad and other tablets that emerged. With iPad and other tablet growth, website developers are being forced to rethink their website designs and functionality to best accommodate the new mobile device medium. Location-based social media applications will present new and exciting opportunities for residents and local governments.

Mobile applications (apps) are a relatively new channel of communication. They may exist and operate entirely outside a local government operation and be separate and apart from any official website. They might take the form of a stand-alone application or one that is fully integrated into a local government's communications center. Gartner predicts that by 2013 mobile phones will overtake PCs as the most common web access device. Add to this prediction that smartphone growth is expected to have double-digit growth in the years to come—coupled with the explosive growth of tablet devices—just as emerging apps will overtake websites. A study by Flurry found iOS and Android apps surpassed 1 billion downloads in the final week of 2011. In addition, according to the analyst firm Berg, 98 billion apps will be downloaded in 2015. These statistics, while amazing on their own, hold either serious threats or opportunities to the public safety community. To date, it's a matter of perspective.

Web-based communications now are being supplemented with stand-alone apps, or apps that appear to be separate and apart from a website. Even outside the Apple App Store, apps have been developed by and for local governments that allow residents to report potholes, animal control issues, graffiti, lighting issues, crimes or accidents, and much more. The new apps allow pictures to be submitted with the latitude and longitude, exact time, and device owner information. Because apps usually are completely separate from a website, there are greater opportunities for innovation and experimentation. This presents the opportunity to develop applications outside the realm of centralized IT, which often can speed up the implementation process. Of course, this very advantage also could create problems if not properly coordinated with the rest of the enterprise.

State and local government leaders and managers are challenged as never before to seek out new and better ways to engage and interact with the public they serve. With a greater than 97% penetration rate of mobile devices, which are becoming smarter with each new model, government leaders must understand and adequately address how the residents they serve prefer to be communicated with and engaged. At the same time, these same managers always must be aware of those in the population who either cannot afford or choose to opt out of communication technology. Today, that number stands around 25% percent, according to the Pew Internet and Family Life Project.

Web 2.0 and civic/social media depends on residents and governments having reliable access to affordable broadband in homes and offices. The Federal Communications Commission's (FCC) National Broadband Plan for America includes a section on public safety communications. While the plan clearly and rightfully addresses the need for greater interoperability and cyber security, it also recognizes the need to leverage broadband technologies to enhance emergency communications *to and from the public.*

The plan goes on to restate the need for the nation's 9-1-1 emergency call system and the emergency alert systems as well as the recognition that Next Generation 9-1-1 (NG911) and Next Generation Emergency Alerting (NG Alerting) technologies are deployed in the near future. Providing sufficient funding to support deployment of NG911 and removing regulatory barriers to its deployment should ensure that NG911 is made available across the country. The report encourages the exploration on how to best enable NG911 and NG Alerting technologies. Since the report was focusing on broadband and less so on applications, it is still interesting to note there was silence when it came to the specific issue of having residents play a more active role.

The National Emergency Number Association (NENA) first identified the need for NG 911 in 2000, when it noted that the current 911 systems cannot meet the ever-growing needs for the future. A set of standards are being developed that basically will shift today's circuit-switched 911 networks to packet-based, or IP-based, thus allowing for a near-seamless integration of voice, video, and data. While basic planning has been under way for some time now, most experts believe that a comprehensive set of national operational standards are a decade away. Another concern voiced by the public safety community regards the ability to fund the purchase of new equipment, personnel, and training.

So, how can governments at all levels best prepare and respond? This remains an open-ended question that public managers at all levels must address with both strategy and action. As we shall see later in this chapter, the good news is that much can be accomplished right now without waiting for a national standard to take hold.

Performance measurement—Measuring success or failure

As public safety agencies plan ahead for social media integration, many leaders seek ways to measure what is working, how, and why. Defining success using social media will take some time to sort out. Today, we have the tools that enable us to measure outputs. We know how many messages go out, how many were opened, and in some cases how long they spent reading or searching an item. We also can measure how many bits and bytes of information we receive, and by whom and how frequently. So, while we can quantify social media results, the softer side or qualitative ratings are more challenging.

With all the growth in local government outreach using social/civic media have come many disappointments. In 2010, the Public Technology Institute (PTI) conducted a national survey directed to those who manage their social media applications, asking about how they were using the largely built-in user metrics. In the past, if a local government were to send out a flyer, the best they could report was the amount of paper printed and, possibly, a secondary reader percentage. New social and civic media, by comparison, contains built-in measurement capabilities for the taking.

Today, when a local government sends a message, it can not only know how many people it sent it to but whether the message was actually delivered or returned, opened or not, for how long, and how long recipients spent on a particular page. The PTI survey revealed that a large majority of managers did not use social media metrics because of three stated reasons:

- They claimed to be too busy and didn't have the time,

- They claimed that it was too complicated and they lacked the proper training, and

- They claimed their supervisors did not care or never asked about social media metrics.

Getting news and emergency information out

During the steady decline in print journalism, some have expressed fear in what they see as a decline in literacy and critical thinking. There is also a growing concentration of media ownership that can have the effect of taking a small news story in one locality and highlight it in such a way that the significance is distorted or perhaps blown out of proportion based on dramatic elements and video. Just as bad, some stories that used to receive coverage have disappeared. Finally, daily news has transformed into daily views.

People are not just reading news online, they are contributing to it. According to the Pew Internet & Family Study "Understanding the Participatory News Consumer" (2010), 37% of Internet users have contributed to the creation of news, commented about it, or disseminated it via postings on social media sites like Facebook or Twitter.

Local governments realize they can no longer rely on traditional media to reach the public. They are developing innovative channels to broadcast their messages through various new media applications, and they have done an exceptional job. Many jurisdictions now offer residents the choice of having specific news and information feeds sent directly to a laptop, home computer, mobile device, etc. Residents can choose to receive public safety alerts, weather and traffic bulletins, or meeting notices. Local governments using new media delivery systems can better control their own need to send out timely and accurate news and information with greater frequency and detail.

Social & civic media implications for public safety

Nowhere has social and civic media gained greater importance than public safety applications. One of the hallmarks of social media is the ability for just about anyone to either create something positive or, in some cases, create havoc. Cybercrime is growing both faster and more dangerous with every passing day. Social media has become an innocent carrier, and despite all the good, it can carry misleading and inaccurate news, information, malware, and theft. It was reported on national news that a family posted a detailed itinerary of their weekend trip and came home to find their house had been completely robbed. Apparently, someone had read the posting and knew the house would be vacant. It also must be pointed out there are many upsides, where local authorities are actively monitoring social media sites to gain intelligence on gang, drug activities, or sexual offenders.

"It used to be we were the first responders!"

Public safety and social media can be reduced to three overall categories: (a) fighting crime, both traditional and cyber; (b) getting the word out or broadcasting critical information to the public; and (c) figuring out ways to use existing resources to encourage the public to send in critical information, including text messages, pictures, or video.

A first-time experience with high-definition (HD) video and sound can be startling. The picture and sound is so life-like that people in other locations can't believe they are not in the same room. Without question, HD Dispatch is here, and many are completely unprepared for its effect on dispatchers and call takers alike. Added to the day-to-day stresses that come along with the job, one now will be subjected to vivid displays of mayhem and violence with high-definition sound. There is a growing need for better training to help prepare staff for the possible horrors that they may soon be seeing and hearing—as if they were actually on the scene.

The consumerization of technology

As noted earlier, residents are moving up to smarter and smarter devices at a staggering rate. However, there are many other technologies that were once available to professionals and are now found on the Internet and in airplane seatback catalogs. The next generation of video cameras will be smaller, lighter, and more powerful in every way. New still cameras can take HD video and can be connected easily to the Internet as well as big screen TVs. Citizens will have forward- and backward-facing higher-resolution cameras that will enable greater peer-to-peer and group-to-group communications. Moreover, just as word processing programs allowed almost anyone to be a publisher, new easy-to-use video processor programs allow almost anyone to be a movie producer.

Now residents can track their kids, spouse, and whomever, with a magnetic mount "hidden" GPS tracking device that can be purchased for under $125. One also can purchase a pen camera with built-in DVR storage for under $100. Finally, one can easily purchase devices that capture every keystroke on a PC or Mac, allowing someone to "snoop" on what emails are being written and to whom, what is being purchased online, what websites are being visited, etc. It may be against the law, but you would never know that by searching for such products online.

Certainly as with any technology, there's always a dark side or vulnerability that needs to be understood. Just two days before the 10th anniversary of the September 11, 2001, terrorist attacks, followers of an NBC News Twitter feed were informed, "Breaking News! Ground Zero has just been attacked. Flight 5736 has crashed into the site. More news as the story develops." Another tweet followed that stated this was not a joke; reporters were rushing to the scene. As it turned out, to the horror of NBC News officials, their Twitter account had been hacked (hijacked might be a better word for it). The news outlet regained control of the account in about 40 minutes. However, for the 112,000 followers, there was much concern, and we know that stories build upon stories, and complete panic could have ensued. NBC Nightly News apologized to the public for three consecutive nights on its national newscast.

One can never be too careful in protecting and monitoring social media accounts. The NBC News Twitter account hacking was considered a "prank" by the proclaimed organizers, and similar pranks are just as likely to affect local governments as respected businesses. They not only compromise trust, they might be used as diversionary tactics by someone or a group with far worse intentions.

Flash mobs, swatting, and gangs

There have been other disturbing examples of people using social media for "swatting" and "flash mobs." Swatting is a term used to describe a situation where someone or groups of individuals e-mail, call, or post an erroneous claim that a particular house or building is being held by hostages. In such cases, a SWAT Team is dispatched, and not only is there a good chance an innocent person may be harmed by the public safety response, it takes special resources away from a place where they may really be needed.

Flash mobs are groups that are called upon to gather at a particular location in a specific point in time. Flash mobs have assembled for snowball fights in parks or to stand completely still in a city's major train station. Public safety officials have been monitoring such activities when they can, but there have been recent accounts where flash mobs have turned into unlawful acts. Groups are dispatched via a social media platform and enter stores, overwhelming staff and stealing everything in sight.

Gangs have found social media to be a convenient tool for communicating among their members, so public safety officials have assigned personnel to monitor gang activity via social media. Social media is always evolving, and new uses are being tried every day—some very good and others very bad. Public safety will be in catch-up mode for years to come.

The rise of the public

What are the implications for local governments? As the private sector is poised to capitalize on the evolving social media revolution, the public will expect no less from their government. Already, the news media is encouraging everyone to be "I reporters" and upload their video clips as stories unfold. Often the first to a scene are amateurs clicking and posting away long before anyone from the media arrives.

Public safety communication officers are beginning to think about new training for their dispatchers, who soon will be viewing emergencies from the field in real-time and in HD. They are concerned that many dispatchers may be unprepared for the potential shock of seeing blood and trauma in ways that could be disturbing, shocking, and far more stressful than a phone conversation.

The new media provides many good opportunities for enhancing communications:

- Streamlined and common communications platform
- Multi-point accessible knowledge bases
- Networked resources
- Real-time data communications and reporting
- Wireless access (multiple paths)
- Multi-media delivery
- Integrated self-service options
- Location-based services
- Cross-agency information sharing
- Potential for improved citizen trust and engagement

With the potential for greater public input, the benefits of social media in the context of public safety really begin to emerge. For example, there is the potential for greater intelligent systems dispersed among a wider population and geographic area(s), mobility and multiple feeds in real-time, and superior technology. One public safety officer commented that many of the built-in cameras found in today's cellphones have better resolution than many of the older fixed surveillance cameras. There are literally thousands, if not millions, of people that have camera phones.

What's more, the public wants to be more involved in making their neighborhoods safer, and they are happily turning to social media to do that. In a recent meeting at police headquarters in the "City of Brotherly Love," Philadelphia, it was reported that four recent murders had been solved through the posting of pictures of possible suspects on social media sites. Working with local police, social media is taking neighborhood watch and community policing to higher levels than ever before. Today, many cities have adopted their own apps that help the general public not only better visualize crime, but encourage citizens to report things that they see or hear.

IT governance in a public safety setting

The traditional boundaries that have existed between public safety IT and municipalities have become further galvanized. The human firewalls that exist serve as an invisible speed bump that can blur the lines of authority—thus preventing greater cooperation and improved coordination. This reminds me of a chief of a police force who unilaterally decided to ditch the department's BlackBerrys in favor of a new smartphone for the force. The municipal IT staff was notified long after the contract was signed, yet they were expected to provide support, training, and a secure server. Needless to say, that did not go over very well; however, examples like that occur each and every day.

Sound IT governance planning is essential to maintain a vibrant and secure set of systems. Social media is no exception, and everyone involved needs to be part of the planning process that includes equipment and software certification, understanding who is in charge of what, and what new skill-sets are needed. As some have said, Web 2.0 is not a wave of the future, but a tidal wave in the making, especially when it comes to public safety.

Public safety has been very effective in using outbound citizen notifications and information that includes Twitter, Facebook, YouTube, Reverse 911™, RSS feeds, special apps, and regular municipal websites. The challenge is how to assimilate incoming communications—photos, videos, text messages, e-mails, or all of them combined. Fortunately, there is technology available that can monitor aggregate messages, and there are artificial intelligence systems that can "scan" and "monitor" all inbound communications and create a real-time situational report about what is happening and where.

Only government can attempt to counter the gap with programs and services aimed at bringing the "have-nots" to the Internet through training, public facilities, and partnerships with other governmental entities and private businesses. The idea that many state and local governments still charge a fee for showing up to a government building for a service instead of going online is misguided and hurts the very people who need help the most.

Anonymous, anonymity, gossip, and rumors (Verify—authenticate)

At the time of this writing, a group calling itself "Anonymous" had taken credit for hacking into a number of prominent business and government websites. As the name implies, this is a group that operates anonymously. The group is skilled and is considered dangerous by government leaders, and eventually the leaders of the group will be caught and prosecuted, but they most likely only will be replaced with another self-styled hacking group. Some are motivated by the thrill of the challenge, while others have more sinister ideals.

Anonymous rumors and pseudo names unfortunately have been somewhat normal in U.S. political affairs. What has changed is that social media makes it that much harder to ascertain just who somebody really is. For local government officials, simply contending with what appears to be anonymous postings, false rumors, and unfounded gossip, can have an extremely negative impact on social order and stability. The environment for the need for trust is as strong as the abundance of distrust one strives to displace. This paradox cannot be ignored, and there has never been a greater time for the need for local governments to establish trust. The need extends to how local governments provide adequate, timely, and truthful information. The term "citizen engagement" needs to become operationalized from a "sounds-good concept" to a meaningful and visible process.

While it may appear that people can manipulate ways to appear anonymous, today's technology also can track and expose those who truly abuse such ruses. It is particularly important to place a burden on those who make public postings, and require respectful language and some form of opinion ownership.

Without appropriate safeguards and policies, local governments may find themselves drowning in a sea of cyber-anarchy. We know that government derives its legitimacy when grounded on trust.

The federal government and the private sector have begun working on ways and standards to better ensure authentication and verification of identities, which is essential to a thriving online environment. In April 2011, the White House issued "The National Strategy for Trusted Identities in Cyberspace–Enhancing Online Choice, Efficiency, Security, and Privacy." This report sets forth a framework at the national levels to begin creating systems that can authenticate and verify online identities, which is a fundamental requirement for our emerging online society.

If you see something, send something

In every airport and train station, there are signs and recorded messages reminding travelers that if they see something suspicious, they should say something to an authority as soon as possible. Often, in today's world, the first to reach the scene of a crime, accident, plane crash, etc., has a smartphone, and they instinctively take and send pictures. Every local and major news outlet encourages this practice. CNN was the first to coin the term

"I-Reporter." When a rare and large earthquake hit the mid-Atlantic region in August 2011, CNN was reporting the event in detail about 20 minutes before local governments began to send out brief messages informing a public of what they already knew before any "official" announcement.

According to a survey conducted by the American Red Cross, more Americans are using social media in times of emergencies. The study also revealed that most Americans who use social media feel they should receive timely assistance from local governments and national response agencies when they post messages in social media. The same people responding to the survey also indicated they still relied on 911 for emergency situations. Nevertheless, 911 was designed for inbound communications, and the interesting finding was the high expectation of actually receiving information in times of emergencies. It should be noted that the study was focused on social media users, which does not account for the nearly one-third of Americans who do not spend time on the Internet.

To be able to make sense of incoming citizen input information, both critical and routine, public safety managers are going to need to have dedicated equipment and staff that do nothing but monitor and respond. Staff need to be as ready to send out alerts as well as handle potentially false and misleading information. Social and civic media systems and platforms cannot be a passive endeavor. They need to be actively engaged and monitored 24/7.

Systems and policies need to be in place that help verify incidents and locations. As public safety agencies at all levels of government continue to collect what is now being referred to as "big data," law enforcement has the challenge of staying in compliance with the law, protecting security as well as privacy, and developing ways to store and analyze data gleaned from multiple sources. We are already beginning to see the rise of artificial intelligence systems that are designed to make sense out of big data. These systems operate in real-time, analyzing and interpreting data in milliseconds that would be humanly impossible otherwise.

Public expectations are on the rise, and public safety managers need to develop new strategies and systems now. Having the ability to accept social/civic media inputs to a command and dispatch centers is an essential new set of tools that can bring about many positive benefits. Residents, by and large, want to make a contribution toward the welfare and safety of their respective communities. They want to weigh in and share photos and basically serve as adjunct surveillance monitors. Harnessing the public leads to greater collective intelligence and can take neighborhood watch to a new and higher level, adding hundreds to thousands or more eyes and ears to the heartbeat of our communities.

DR. ALAN R. SHARK is the Executive Director and CEO of Public Technology Institute (PTI). Celebrating its 40th year, PTI is a national, non-profit organization that focuses on technology issues that impact local government and thought-leaders in the public sector.

Dr. Shark's career has spanned over 28 years as a highly recognized leader in both the non-profit management and technology fields, with an emphasis on technology applications for business and government. He is an associate professor of practice at Rutgers University where he teaches a masters-level course on technology and public administration.

He is author of the book Seven Trends That Will Transform Local Government Through Technology *(2012), coauthor of* Web 2.0 Civic Media in Action *(2011), and an author and executive editor of the book* CIO Leadership for State Governments: Emerging Trends and Practices *(Spring 2011). Dr. Shark also was executive editor of the book* CIO Leadership for Cities and Counties: Emerging Trends and Practices *(2009) as well as* Beyond e-Government: Measuring Performance *(2010) and* Beyond e-Government & e-Democracy: A Global Perspective *(2007).*

Dr. Shark was elected a Fellow of the National Academy of Public Administration. He received his doctorate in public administration from the University of Southern California.

Chapter 9: IT Governance in Public Safety

BY DAVID WHICKER

Introduction

The role of the CIO seems to be ever evolving, whether that role is in the private or public sector. CIOs in the government sector are tasked with understanding and supporting a great deal of public services, which can span across multiple disciplines and agencies. More specifically, when leading in local government, public safety can pose a number of real challenges as well as opportunities. For years now, we have been bombarded with facts, studies, and information regarding IT governance until the word governance itself seems like another buzz word. Sometimes we can get so caught up in frameworks, models, and studies that we create for ourselves an environment of smoke and mirrors. I, however, would like to submit to you that IT governance should not be and does not have to be this challenging.

I spent the majority of my career in the private sector, and I have seen governance, or at least the attempts of governance, up close and personal. Based on my exposure, I am convinced that there is no one correct approach, no one specific model or perfect structure. Frameworks and guidelines can be very beneficial; however, at the end of the day, you must find what works best for your team and your organization.

Role of the CIO in public safety

As public sector CIOs, we have to be very familiar with how significant our roles are in our organizations, especially the effect we can have on public safety. As the technology leaders, we are charged with ensuring we meet the objectives of our organizations through business alignment. Today's CIO must have a broad skill set outside of just wearing our engineer and support hats. A CIO must be able to market and sell information technology and the services their respective departments provide throughout the organization. The CIO must be somewhat proficient and continually develop the following areas:

- Marketing/sales
- Customer service
- Information technology
- Strategic planning

At some point in every IT professional's career, he or she has to develop an answer to the question, "What do I want to be when I grow up?" I am a firm believer that there is a clear progression as an IT professional, and that if ultimately one decides he or she wants to become a leader, they must follow the path shown in Figure 1. The journey to becoming a leader, the paths, and the amount of time it might take varies and would require more than one chapter, so I would like to take a basic approach and provide what I feel are high-level but beneficial concepts. One can progress rapidly or slowly to becoming a leader, but the amount of time it takes is not as important as the fact that you recognize it is a journey, there is a logical progression, and there is no final destination.

Figure 1.

Governance and politics

Politics in the private sector was always a looming topic and somewhat comical now that I work in the public sector and it actually has meaning. The fact of the matter is that public sector CIOs face a diffuse power structure in that there is not always a top-down approach or typical chain of command you would find in the private sector.

The nature of working in the public sector means you will have to deal with politics and bureaucracy consisting of elected officials, appointed personnel and all types of established boards. In the realm of public safety, it only gets more cumbersome as agencies such as 911, emergency services, and law enforcement are a tight knit community that almost speaks a different language, much like the acronyms we throw around in IT. With public safety, you can almost rest assured you will come in contact with all types of governing boards, councils with titles such as Fire Chiefs Council, Emergency Services, and Rescue Board. The list can go on. Although these structures were designed with accountability and governance in mind, it sometimes can be uncharted waters for the typical CIO and can cause some CIOs to wave the white flag.

Let me offer some simple advice based on my experiences in dealing with the various boards, elected officials, appointed personnel, and the like:

- Communication is key. The more routine information you can provide, the better. I recommend routine visits, walk-throughs, meeting attendance, and socials for starters.

- Staying as neutral as possible always tends to be the best route. Simply put, your role is to provide support to the organization and the position regardless of who holds the title.

- Creating partnerships regardless of the structure will ensure or maximize success in whatever governance framework you choose. Partnerships built on trust will trump any official agreement or policy period.

Achieving business alignment through vision and strategy

Business alignment and governance are synonymous almost in that one of the most important reasons we seek IT governance is to reach the utopia of alignment between the business and IT. We often talk about business alignment in IT, but I think it's important to have a working definition. My definition of business alignment is simply this: using information technology as part of a solution to achieve any number of desired business goals and/or outcomes.

As we consider IT governance as it relates to public safety, vision and strategy are two major components that deserve further attention, as they both play a critical role in the success of the CIO. Regardless of the organizational structure, you cannot lead as effectively in information technology as the CIO if you do not include them in your overall strategy and reach out to the public safety agencies you support either directly or indirectly.

VISION

I believe Andy Stanley said it best: "Vision is a clear mental picture of what could be, fueled by the conviction that it should be." When I look at leaders, in general, I can tell within the first few moments of meeting them whether they have "what it takes" by the passion in their words. I, for sure, am not the final judge on who is a leader and who is not, but I always can spot passion and conviction. When someone has a burning desire to upset the status quo because they know and can see what could be, there is your leader.

CIOs today must have vision. You cannot "move to the next level," "get to where you are going," or "drive change" if you have no clear destination that can be articulated to others.

STRATEGY

I think we all remember the days when our power was found in being able to fix things and knowing the insides—a scary thing to those whose job it is to trust you. At some point we must realize that the best decision model and approach is a shared decision model with the business, especially in public safety. Learning the business and processes throughout different departments and business units is one thing, but public safety comes with its own set of challenges.

Often, as a public sector CIO, it can be difficult to create a strategy that falls in the category of business alignment because a strategic plan or direction may not exist for your city or county. To ultimately achieve business alignment through any use of governance, the logical progression of thought would be to ask, "What business are we in?" In the case of public safety, it's saving lives; protecting and serving; and providing civil services, 911 communications, and emergency response. Once you understand the critical nature of this, you can begin to think big picture and partner with those agencies to move in a unified strategic direction.

I would like to submit to you steps that can assist you in moving toward business alignment through strategic planning:

- **Start from where you are no matter what that means.** You have to start from somewhere, or you will go nowhere. Zig Ziglar says, "When you aim for nothing, you will hit it every time." Get a baseline understanding and assessment of what information technology, policies, procedures, and methods are in place today. Document those well in the form of a SWOT (strengths, weaknesses, opportunities, threats) analysis, and don't attempt to hide the findings no matter how uncomfortable. You may be surprised at an individual's reaction and just how much they already suspect may or may not be correct.

- **Begin taking a strategic look at the present and future with a firm understanding of where you've been.** Once you have this foundation built, you can begin to think clearly. When all cards are on the table, it is much easier to be honest about what you can provide now and what you would like to provide. This becomes a key component of your overall strategy.

- **Break ground (if you haven't already) with the leaders, boards, directors, chiefs, managers, councils, and whomever else you need to bring to the table, and begin communicating openly, honestly, and candidly.** Sometimes the initial step is letting everyone know you are here to help, you want to listen, and you want to use your knowledge, talents, and abilities to provide them with information technology as a vehicle to improve their everyday efforts in serving residents.

- **Develop goals and objectives that are win-win.** I am sure this sounds easier than it appears, but I truly believe there is always a middle ground in which policies, procedures, and standards do not conflict with the business needs of an agency or department. Find a way to reach "yes" without compromise. With the right attitude and tone, this is not unreasonable.

- **Create a living document in the form of a strategic plan.** There are more than enough examples on the web and in textbooks to guide you in your efforts. The main thing is that you openly and honestly provide a strategic direction where the details can be flexible when needed to meet the needs of the organization.

The diagram in Figure 2 simply displays the logical progression but ever-evolving process of how your strategic plan moves along with the organization.

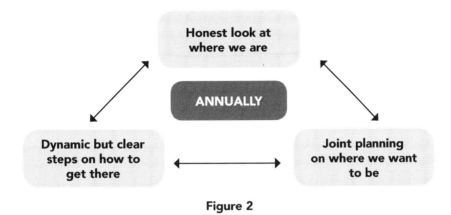

Figure 2

When developing a strategic plan for information technology in your respective organization, ideally you would want to build off an existing strategic direction or plan. Unfortunately, you may find that this does not exist for any number of different reasons. Don't get discouraged if this is the case, and again start from where you are. If this plan does not exist, you still move forward as a leader with your strategic plan and move down the road of executive leadership buy-in.

Leadership principles never change

Principles always intrigue me because they generally hold true regardless of any circumstance, and you cannot escape them. Over the course of my career and years in studying leadership, I have found that all great leaders tend to have similar characteristics in common. I am not entirely convinced that a formula can be given to develop these characteristics, but whether inherent, learned, or developed over time, these characteristics always are present:

- An understanding that leadership is a stewardship;

- An understanding that leaders are ultimately accountable;

- An understanding that leaders are held to a very high standard;

- An understanding that leaders set the tone, mood, and direction of their organizations;

- An understanding that leaders have the option to approach every fire with a bucket of gasoline or water;

- An understanding that regardless of the outcome, leaders have to make a decision and stand by it but be willing to admit it may not have been the right approach;

- An understanding and keen awareness that being a leader means that you will have to stand alone at times but making tough decisions comes with the territory.

At the end of the day, influence is what matters, not control. We can have policies, procedures, chain of command, and the perfect framework and execution of IT governance, but without influence, it's just not enough. With public safety and the sometimes political nature of board/council-controlled environments, we ultimately have no real control even if we desired it. The fact of the matter is that control really is not needed when you have true influence, which always has the underlying trust and respect of our fellow executive leaders.

We have to get to a place where IT is not perceived as an obstacle or barrier to progress but that all parties see us as trusted and innovative, which means we prove that we truly care and understand the daily business needs and challenges that face individuals serving in public safety.

Centralization and consolidation

A centralized IT model

There are several schools of thought and scores of books written on what centralized and decentralized IT models look like and how to effectively implement either approach in an organization. Fortunately for us, this section is not another one of those. In regards to IT governance and public safety, the model is at least worth briefly reviewing because serious decisions must be made that are not always solely left to the CIO, and serious lasting implications must be considered.

It is not always possible to have a completely centralized model, assuming that would be your aim. In fact, there can be different approaches dependent on several circumstances that the CIO may have no real authoritative control over. Some CIOs may find that they simply must settle for a completely decentralized model, or at best a hybrid approach. In the most extreme circumstances, the CIO will find him/herself in the terrible situation of no control or involvement at all.

The following are three typical models in public safety and IT governance in local government:

- Centralized
- Decentralized
- Hybrid

Centralized

In a centralized IT model or approach, all IT services and support (enterprise, business, application, service desk) are provided by a central service provider—typically the organization's IT department.

Decentralized

In a decentralized IT model or approach, one or more business units may house their own IT resources, which could look any number of different ways. More specifically, in my experience with cities and counties in North Carolina, law enforcement and 911 might have their own IT personnel providing help desk support, networks, and telecommunications.

Hybrid

Regardless of budget allocation, direct reporting structures, and/or department FTE counts, a hybrid model consists of joint communication and cooperation between IT and the business units to use all available resources to ensure the respective agency or department gets the information technology attention needed to succeed. If a picture is worth a thousand words, then the diagram shown in Figure 3 illustrates the general idea.

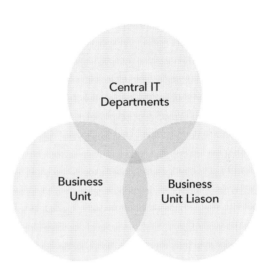

Figure 3.

The Business Unit Liaison Resource Associate...

It amazes me how creative we can become with titles and positions. I often marvel in what we might come up with next when attempting to create "synergies" between IT and the business units. The fact is there can be tremendous value in implementing what I simply call a liaison. The most effective individuals I have found in both the private and public sector are those who have the unique ability to translate between IT and the business to achieve desired results.

To centralize or not centralize ... that is the question

I am not sure there is one exact model that could be considered the "cookie cutter" answer. One also has to take into account that the decision to centralize may or may not be the CIOs to make or even include the CIO for that matter. Ideally, the CIO would at the very least have a seat at the table for the discussion, but in some cases that could take a little effort to achieve.

I personally believe that, whether internal or external, all IT services should be provided and orchestrated by a central IT department, which includes a very grey area of providing business analysis or business resources to the business units. However, my approach often would fall in the category of "hybrid," depending on a number of different circumstances.

To better illustrate and assist in the decision-making process, Figure 4 and Figure 5 outline some of the advantages/disadvantages as well as general advice on how to approach each model, assuming you reach a decision point.

Leadership advice for each model		
Centralized	**Decentralized**	**Hybrid**
• Protect IT's reputation in the organization by providing a robust infrastructure, solid service desk and any other basic IT utility service • Over communicate throughout the organization on the happenings of IT • Ensure a solid IT staff and develop a unified front	• Find your influencers and decision makers in the organization, and focus on buy-in from that group • Ensure solid decision-making matrix in place • Push standardization and have solid policies and procedures in place	• Ensure solid communication/relations with business unit liaisons and/or IT personnel if staffed outside central IT • Request "dotted line" reporting structure to provide high-level direction and initiatives • Push for centralized IT utility-type services (service desk, network support, infrastructure, telecom, etc.)

Figure 4.

Advantages/Disadvantages of Centralized/Decentralized		
	Advantages	**Disadvantages**
Centralized	• Central authority and governance • Logical chain of command • Can be geographically independent • Single location of support • Central location for standards/policies/procedures • High system availability • Tendency to lean toward a one-size-fits-all approach	• Individual departments/agencies may not get the attention or focus they feel they should • Can limit adaptability to specific needs of specific business units
Decentralized	• Responsiveness • Focus/Clear priorities • Argument for greater flexibility • More business unit and end user support leads to potentially greater satisfaction • Potential for quicker response times and/or attention	• Unclear lines of demarcation • Duplication of resources/efforts • Could lead to non-standardized environments • Potential power struggles or tension on direction of organization IT and specific agency IT • Lack of accountability to knowledge experts • Potential to lose strategic alignment if not managed appropriately

Figure 5.

IT support in public safety

When lives are at stake, the word "emergency" takes on an entirely new meaning. A common rant as an IT professional has always been that to the end user, "everything is an emergency." How often have we gotten upset over those after-hours calls we received that were deemed by the user to be emergencies, when, in reality, checking their state retirement account balance easily could have waited until the next business day. When you find yourself responsible for an IT organization that directly supports public safety operations, such as 911 call centers, it can be a little more interesting and sobering to say the least.

There are solid utility-type services, processes, or procedures that you must have in place to even begin to successfully support a public safety agency, and the moment you sign yourself up for that support, you may be knee deep in some serious trouble without them:

1. A robust infrastructure.

2. A thorough documented service desk procedure both during normal business hours and for after-hours emergencies.

3. A solid Service Level Agreement (SLA) with the respective agency.

4. The appropriate redundancies in support coverage, infrastructure, and enterprise applications.

From Technical Support to Business Analyst…get out of the office

IT organizations are finding today that their roles are shifting from primarily providing the infrastructure and support they are traditionally accustomed to and requiring more business process support roles and understanding. In fact this is so desired and needed by all business units, especially public safety that you will either fulfill this role or someone else will.

Business units, agencies, and departments are looking for translation of information technology in terms they can understand. A paramedic trying to document a patient's information, health stats and performed procedures to send to the emergency room before arriving, really doesn't have time to understand technical jargon or mess with bells and whistles they know nothing about. We can have the best, brightest, and most educated computer scientist in our IT departments, but if they can't translate a real business need into a solid solution, then we are extremely ineffective and will soon, if not already, have a huge problem on our hands.

I often am amazed at how difficult it becomes for IT professionals in general to share their knowledge and effectively communicate with those around them. We far too often stand back, joke, and complain about the way others go about performing their routine job duties so inefficiently when we hold the very keys to unlocking that efficiency. A word of advice to you and your IT teams: if you find yourself heavy on the PC technician "break/fix" reactive side of information technology, you may soon find yourself looking for employment elsewhere.

The line between city/county infrastructure and public safety infrastructure

Often, one of the struggles we face in supporting public safety is simply determining where to draw the line in meshing the city/county systems and underlying infrastructure with those of public safety. This is one of the primary reasons we need a solid IT governance structure in place. Let me just say, though, there can be a balance or even a shift to one side or the other. I believe that as CIOs, we can provide a solid support model that incorporates third-party services, shared infrastructure, and redundancies without compromising public safety needs and requirements in the process.

Asking yourself the following questions when considering how to proceed in this area may prove extremely beneficial in making your final decision:

- What funding sources are involved, and what are the legal and statutory requirements?

- What are the risks in a shared infrastructure, especially one that is managed, monitored, and maintained by the IT organization?

- Is there a balance of third-party systems, support, cloud services, or oversight that could ensure the appropriate requirements are met for the respective public safety agency?

- Does the IT organization I lead have the skill set, knowledge, response time, and capabilities to successfully support public safety needs 100% of the time?

Consolidation: City/county services in action

I have had the privilege of participating in a county-wide initiative to consolidate 911 PSAP/call centers in Rockingham County, North Carolina. As of this writing, the consolidation project is in the initial stages of implementation being directed by a recently established "911 Governance Board." Together, elected and non-elected representatives from the cities, towns, and county came together and applied for a PSAP Consolidation Grant, which resulted in an award of $7.826 million to Rockingham County by the North Carolina 911 Board.

The purpose of the grant award and resulting project is to consolidate current 911 PSAPs between Rockingham County 911, Reidsville Police Department, and the Eden Police Department. The consolidation will result in dispatch for all law enforcement, fire departments, emergency medical services, and rescue squads for the cities/towns of Reidsville, Eden, Madison, Mayodan, Stoneville, and Wentworth. Rockingham County accepted the grant on December 5, 2011, and the project has a projected completion date of December 5, 2014.

A governance board has been established by an inter-local agreement to oversee the implementation and oversight of the new center titled, "The Rockingham County Emergency Communications Governance Board." The inter-local agreement is composed of seven voting members and two non-voting members:

Voting Members

- Two police chiefs
- Rockingham County Sheriff
- Representative of the Rockingham County Fire Chiefs' Council
- Representative of the Rockingham County EMS/Rescue
- One citizen representative nominated by municipalities participating in the inter-local agreement
- One citizen representative nominated by the Rockingham County Board of Commissioners

Non-Voting Members

- Rockingham County 911 Director
- Rockingham County CIO

All Rockingham County Emergency Communications Governance Board members were approved by the Rockingham County Commissioners to ensure fair representation to all demographics and geographic areas of Rockingham County.

I am a proponent of consolidating duplicative services between governments no matter what the scale whenever possible for efficiency and effectiveness for the citizens we serve. More often than

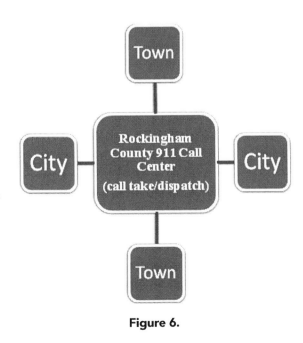

Figure 6.

not, the bureaucratic systems and political structures we have in place tend to prevent these types of collaborative efforts. I am very proud to have had the opportunity to be a part of a conceptual model that now is unfolding daily and by all accounts is expected to be a success for Rockingham County residents.

Conclusion

The role of the CIO in regards to public safety is one that is not to be taken lightly. Regardless of the governance model and organizational structure, you must have a seat at the table when information technology and public safety meet. There are real challenges today faced in our local government organizations involving funding, resources, and an ever-changing political environment that cannot be ignored. We must find a way to use our talents and abilities to offer bold decisions, keeping in mind that the technology solutions we provide reach far beyond the walls of our department and organizations regardless if we or others only consider ourselves as internal service departments.

In closing, let me briefly say that I understand the difficulties and struggles facing the CIO daily that cannot be covered in one chapter in a book. For those either aspiring to become a CIO or who currently serve as the CIO in their organizations, please remember that you can make a difference; your leadership is a stewardship that should be viewed as a privilege and is only temporary.

DAVID WHICKER, CIO for Rockingham County, has 11+ years of IT experience in both the private and public sector. Whicker joined Rockingham County in February 2010 after serving as a Systems Engineer for MillerCoors. The bulk of his career has been spent in the private sector serving in areas of project management, network design and systems engineering. Whicker is currently enrolled in the NCGCIO program at UNC Chapel Hill and holds a Bachelor of Science in Computer Information Systems from High Point University. Recent projects with the county include project oversight, network design, and relocation of resources to the county's new LEED-certified Judicial Center; migration of legacy email systems to Google Apps; redesign of the county's network and cabling infrastructure; virtualization of the county's servers and implementation of thin computing using Citrix XenDesktop for VDI. As a sitting member on the Rockingham County Emergency Communications Governance Board, Whicker leads all IT consolidation and planning efforts regarding the development of the new 911 call center. David is passionate about leadership in information technology and process improvement primarily in local government agencies.

Chapter 10: Innovation in the Homeland Security Enterprise—Ensuring New Technologies Meet End User Needs

BY BRAD PANTUCK

The purpose of this chapter is to recommend a strategic approach for federal Research and Development (R&D) Programs in the Homeland Security Enterprise (HSE) to effectively and efficiently partner with their end users to research and develop technologies, protocols, guidance documents, and standards that will ultimately better protect lives and property. The HSE includes federal, state, local, tribal, territorial, nongovernmental, and private-sector entities, as well as individuals, families, and communities who share a common national interest in the safety and security of America and the American population (U.S. Department of Homeland Security [DHS], 2010). The chapter prescribes quick delivery of prototypes and a deep understanding of end users' operational realities so that technologies can be easily and meaningfully integrated by end users.

The first section of the chapter identifies the challenges HSE R&D programs typically face in order to innovate; the second section proposes a user-centered design methodology to effectively meet those challenges in a way that provides the "boots-on-the-ground" practitioners with resources to enhance their preparedness and response capabilities.

Homeland security R&D challenges

Many homeland security R&D solutions take years to develop and are unsuccessfully integrated into the end users' operations (GAO, 2008). This is due to a number of inherent challenges to developing and deploying valuable and innovative solutions, including:

- Managing long development cycles, changing needs, and time constraints: Too often, federal R&D projects take years to yield meaningful solutions and risk creating products that don't meet end user needs due to the long lag time between concept and delivery. The needs of the end users can change over time. If the R&D timeframe is long, engineers may continue to work on a solution that has become obsolete. While the exploratory nature of R&D makes it difficult to predict milestones and timelines, greater project management is necessary to deliver valuable products in a timely fashion.

- Defining scope and end users: There are a number of different challenges facing the emergency response community and limited resources with which to address them. A federal agency must determine the scope of activities it will address and, within that, define the end users. This can be difficult because of the complex network of individuals involved in the HSE. Are we targeting IT managers or practitioners in the field? GIS programmers or emergency operations centers? State-level policy makers or emergency services departments? Defining the end users can be particularly difficult in the development of guidance documents, where there is a temptation to target all possible audiences—even when doing so may mean no group is sufficiently served.

- Accessing appropriate end users: Once end users have been identified, the federal government faces the strategic and administrative challenge of effectively accessing that community for their input. With approximately 100,000 public safety agencies across the country (COMCARE, 2007), the HSE needs a systematic way to gather input directly from the end-user community versus relying on DHS component agencies to represent the needs of the local emergency response community.

- Gathering requirements for technologies that are new and innovative: Even if the HSE has identified the correct end users, the traditional requirements-gathering approach does not work when the technology is completely new. Imagine in 1880 asking a typical citizen, accustomed to riding a horse from here to there, about his or her requirements for an automobile. At the beginning, the end users may not be able to explain the problem or realize they have a problem at all.

- Avoiding the "technology push" pitfall: The HSE partners with talented scientists and engineers who excel at researching, inventing, and addressing technical problems. They have an innate curiosity to "solve the problem in front of them" but can become entrenched in the theoretical possibilities of their work, focusing on "what's possible" rather than what would be most useful to the end users. Despite their technical savvy, they are often insulated from the realities of the "boots-on-the-ground" emergency responders and instead develop solutions based on their own assumptions of what would be useful. This problem is reflected when the engineers must "market" their technology to emergency responders rather than the solution having been requested by emergency responders in the first place.

- Transitioning R&D to the end users: While the HSE does develop prototypes, it relies on others to make them operational. That is, the federal government often goes through the complicated and time-consuming process of appropriately engaging and enlisting industry, associations, federal partners, and others to help build and transition the products in a fashion that emergency responders can easily adopt into their operations.

User-centered design approach

A user-centric approach addresses these challenges by engaging end users in a meaningful way throughout the development process. Figure 1 depicts an approach in which end users are involved in the research and development process throughout the product development lifecycle. A single, diverse group of executive-level emergency responder representatives defines the major problem areas and guides the federal government to potential pilot projects. Once a pilot project is chosen, the federal researchers partner with a specific group of end users to co-design inexpensive, sequential prototype solutions with the end users based on iterative feedback throughout the process.

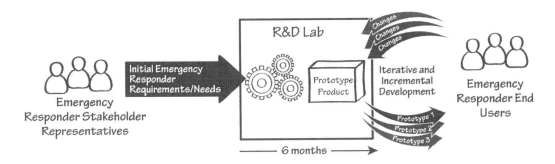

Figure 1. Iterative, User-Centered Design Approach

By breaking down the barriers between the engineers at the federal level of government and the end users, the HSE is able to ground R&D efforts in customers' day-to-day operations, taking into account the subtleties of their environment and tasks. The result is that technologies can be immediately integrated into end users' operations.

In contrast, are the "Technology Push" and "Traditional Requirements-based" approaches, both of which have significant shortcomings. See Figures 2 and 3.

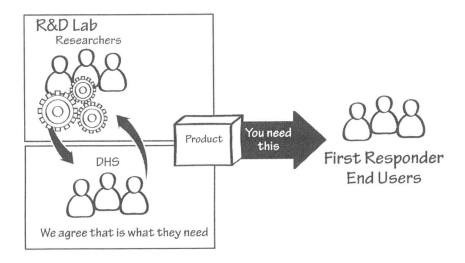

Figure 2. Technology Push Approach

"Technology Push" describes an approach in which agencies identify needs and opportunities, conduct R&D internally, and try to "market" or "sell" their product to the end users, hoping it meets their needs. This approach is employed in many R&D efforts and is often based on "what's possible" or what is most valuable to the federal labs or leaders of the initiative. This approach does not require extensive stakeholder engagement, and in many cases, solutions take years to develop before they are available to the end users. CIOs or R&D managers who allow for this approach risk making large investments that are not tied to end user needs.

Likewise, when the end users are only involved at the beginning of the development cycle, as in the "Traditional Requirements-based" approach (see Figure 3), products evolve based on the developers' assumptions and an incomplete picture of the end users' operations. Such solutions have a hard time being integrated in to an agency's operations. If they do become adopted by end users, it's usually after costly corrections to the product.

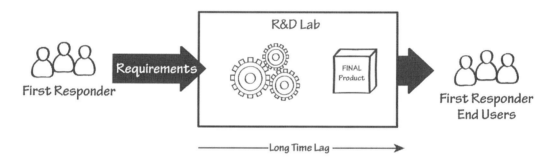

Figure 3. Traditional, Requirements-based Approach

User-centered design methodology

Before engaging in user-centered design, it is important for a federal HSE R&D program to define the high-level priorities of its stakeholders. This should be done in conjunction with executive-level representatives from the end-user community. Before engaging with the end users to solve the details of the problems, the executive-level representatives need to define the high-level problem areas. Laying this strong strategic foundation will allow the program to scope the issue, set up a roadmap, and better define which users should be involved in the design process. Therefore, the following initial activities are recommended:

- Stakeholder interaction: Support and facilitate interaction with a range of relevant stakeholders, including industry, to broaden the knowledge base and opportunity for information sharing among execution team participants and within projects. For existing programs, carefully examine the stakeholder community, including who is currently engaged in the federal HSE R&D program's efforts and, most importantly, who is not engaged. Perform an analysis of why it is important to work with the program's current stakeholders and what specific groups are missing. Leverage the outreach efforts of the existing stakeholders to diversify participation, as needed, to broaden and strengthen the breadth of stakeholder resources.

- Define the federal HSE R&D program's domain: Use the federal HSE R&D program's vision to define what it does and who it serves. This will help to articulate its domain. This domain clarity should be shared internally and externally, and will help to scope the appropriate projects for the execution team.

- Identify the future state and develop a roadmap: Evaluate and define the current state of the federal HSE R&D program's domain, including tasks it works on that fall within this domain. Leverage internal and external stakeholders to identify an ideal future state of the domain. After recognizing the possible barriers to reach this future state, outline and document the necessary initiatives to achieve this ideal future state. Use these initiatives as a roadmap for the execution team. Monitor and upgrade the roadmap as necessary.

Executive-level stakeholder engagement case example: Commercial Mobile Alert Service (CMAS) Forum

The CMAS Forum was hosted by a federal HSE R&D program in July 2009 to bring relevant CMAS stakeholders, including message originators, emergency responder organizations, industry organizations, academia, and organizations representing special needs populations, together to determine the scope, vision, and clear, tangible next steps for the federal HSE R&D program. CMAS Forum participants worked together to identify needs for the federal HSE R&D program to address and define clear and actionable next steps for stakeholder involvement, and to define potential pilot projects.

Once stakeholder priorities or requirements have been identified, the user-centered design approach outlined below can be used to quickly develop products and solutions that genuinely meet the needs of the end-user community:

1. Build multi-disciplinary execution teams to ensure meaningful input from the end-user community, partner to identify a strategic path forward, and manage and develop the necessary solutions (Foster, 1986). These teams should include federal program managers, representatives of the end-user community, project-specific experts (e.g., technology, domain process, facilitation), and transition partners, as needed. By establishing multi-disciplinary execution teams with a defined set of structural and management activities, the teams will be best positioned to collaborate and execute on solutions for the emergency responder community.

Multi-disciplinary execution team case example: Virtual USA

As geospatial visualization tools became prevalent in state emergency operation centers (EOCs), the states needed a way to quickly share information across state borders. The

HSE co-developed a prototype solution with the EOC operators and technicians to enable users to share what they wanted with whom they wanted for however long they wanted. Operators identified what data is typically required for decision makers, technicians knew how to web- and geo-enable data feeds. Short-term pilots produced a new version of the prototype every six months, rapidly integrating feedback from the previous version.

2. Immerse the multi-disciplinary teams in the day-to-day operations of the technical and operational end users. Asking end users about their requirements is a great way to learn what is important, but allowing observers to genuinely experience the end users' operational realities will lead to a richer understanding of their latent needs (Amabile and Whitney, 1997).

Team immersion case example: Integrated Public Alert and Warning System (IPAWS) requirements gathering

In order to understand the current and future alerts and warnings system needs and requirements of the state and local emergency management community, a federal HSE R&D program sent a set of teams to emergency operation centers across the country to observe and interview emergency managers in rural, urban, and suburban communities. These hands-on experiences provided a more accurate view of the emergency managers' day-to-day needs and allowed the federal government to more effectively plan for the deployment of IPAWS.

3. Co-develop solutions with the emergency response operators and technicians. End users' operational expertise and engineers' understanding of what is technically possible are a powerful combination (Utterback, 1996). Designing solutions with the end users ensures that the technology will be on target, and it leads to greater buy-in once the prototype solution is delivered.

Co-developing solutions case example: Interoperability Continuum

Many years ago, it was widely believed that the problems associated with radio interoperability were mainly technical: Different proprietary radios could not talk to each other. After interviewing dozens of practitioners, the federal HSE R&D program team co-developed the Interoperability Continuum with the emergency response community. The Interoperability Continuum is a guidance document that changed the way the public safety community thought about the problem. It teaches the community that radio interoperability has five components: governance, standard operating procedures (SOP), technology, training and exercises, and usage.

4. Quickly build prototypes and get them into the hands of the end users for feedback. A rapid, iterative process of product development, validation, and revision enables the federal HSE R&D program to quickly operationalize solutions where there is success or "fail fast" trying (Leanard, 1995). Prototypes developed using this pro-

cess are more likely to take into account the end users' operational realities and ultimately will be more widely adopted. For this approach to work, the HSE's engineers must be willing to deliver imperfect solutions.

Rapid prototyping case example: Multi-band radios

The HSE conducted lab testing, short-term demonstrations, and pilot projects as part of its testing and evaluation of multi-band radio prototypes. Following short-term demonstrations with fire recruits, the HSE's performer received feedback that the emergency call button was not accessible enough for firefighters wearing gloves due to its small size and recessed location between the volume knob and antenna. The performer was able to quickly adjust the size and positioning of the emergency call button and improve the functionality of the product prior to additional practitioner testing.

Conclusion

This chapter argues in favor of adopting a user-centric approach, engaging end users in a meaningful way throughout the development process, in the federal HSE R&D programs. By breaking down the barriers between the engineers at the federal level of government and the end users, as well as grounding R&D efforts in customers' day-to-day operations, the HSE will be able to quickly co-design prototype technologies and knowledge products that "boots on the ground" practitioners will use.

BRAD PANTUCK is a Technology Transition Manager for RAE, LLC, an Arlington, Va.-based firm that helps assess new technologies for U.S. Navy use. He advises the Office of Naval Research on which military technologies have value for the domestic public safety community. Mr. Pantuck has extensive experience in homeland security R&D, and specializes in commercialization, patents, and technology transfer. This chapter was written with colleagues during his tenure with Touchstone Consulting Group, a Washington, D.C.-based management consultancy. With Lindsey Gualdoni Touchstone Consulting Group.

REFERENCES

1. Amabile, T. and Whitney, D., Corporate New Ventures at Proctor & Gamble, Harvard Business Review 9-897-088 (June 20, 1997): Pages 11-12.

2. Ervin, Clark K. and Aylward, David K., Next Generation Inter-organizational Emergency Communications: Making Tangible Progress While Broader Efforts Continue. COMCARE, 2007. Page 1.

3. Foster, Richard. Innovation: The Attacker's Advantage. New York, New York: Summit Books, 1986. Pages 54-57.

4. Leanard, Dorothy. Wellsprings of Knowledge. Boston, Massachusetts: Harvard Business School Press, 1995. Pages 111-134.

5. Stana, Richard M. U.S. Government Accountability Office. Testimony Before the Committee on Homeland Security, House of Representatives Secure Border Initiative Observations on Deployment Challenges. September 2008.

6. U.S. Department of Homeland Security Quadrennial Homeland Security Review Report: A Strategic Framework for a Secure Homeland, February 2010, Page iii.

7. Utterback, James M. Mastering the Dynamics of Innovation. Boston, Massachusetts: Harvard Business Press, 1996. Pages 160-163, 179-180.

Chapter 11: The Nationwide Public Safety Wireless Broadband Network

BY BILL SCHRIER

On February 22, 2012, the United States Congress passed the Middle Class Tax Relief and Jobs Act of 2012. A provision of that law, generally overlooked by the national news media and most of the nation, will vastly alter the way the 300 million people of the United States are protected and served by their public safety and other government agencies.

The law allocates $7 billion from spectrum sales[1] over a period of 10 years to build a Nationwide Public Safety wireless Broadband Network (NPSBN).[2] The law directs the Federal Communications Commission (FCC) to create interoperability standards for the network and directs the Department of Commerce to create a new First Responder Network Authority (FirstNet) to build the network.

The Spectrum Act sets aside 20 megahertz (MHz)[3] of valuable spectrum in the 700 MHz band[4] for the exclusive use of the NPSBN. It also directs the FCC to license that spectrum to FirstNet, which will then control how the network is constructed.

The administrative structures and processes to build the NPSBN will be clarified by the end of 2012. The construction of the network will probably take the full 10 years. And the financing depends upon freeing and selling spectrum, which is a difficult, multi-year, FCC initiative.

This chapter discusses two topics: first, the need for innovative approaches to building and operating the network, and, second, how the network will change public safety in the United States.

Figure 1. The logo of the 21 cities, regions, and states that are building the first parts of the NPSBN.

Traditional public safety communications

For almost 100 years, radio has been used to dispatch police officers and firefighters to the scene of crimes, fires and medical emergencies. Most cities, counties, and states build their own land-mobile radio (LMR) radio networks, with their own tower sites, radio switches, electronics, mobile radios in vehicles, and handheld radios used by individual officers. LMR networks are now extensively used in government not just for dispatching police and firefighters, but also transportation workers, public works departments, transit systems, and electric, water and gas utilities. In this widely used traditional model, jurisdictions and agencies build, own, and operate all the facilities associated with such LMR networks.

The traditional model has a number of problems:

- It takes a lot of money build LMR networks. King County (Seattle), Washington, spent $62 million in 1992-1995 to build a network of 26 radio sites with about 15,000 handheld and mobile radios serving fire, police, and public works departments protecting a population of about 2 million people.

- The large capital cost means most jurisdictions must ask voters to approve special taxes or levies to fund the construction. This is an uncertain form of funding.

- Most cities, counties, states, and special districts have boom-and-bust budget cycles. When the general economy is good, their revenues are good. When there is a recession or economic contraction, demand for most services increases while budgets are slashed. This cycle often means these jurisdictions that build and operate public safety LMR networks do not have adequate funding to maintain, staff, and upgrade the networks.

- The concept of "public safety" has expanded. In addition to traditional "first responders"—law enforcement, firefighting and emergency medical services (EMS)—"public safety" now includes electric/gas/water utilities, transportation and transit departments, public health, and similar functions. As just one example, electricity is vitally important to our high-tech society and to the public health and safety of everyone, so the role of electric utilities has a distinct "public safety" flavor. These other agencies are sometimes called "second responders." But most of these second responder agencies build and operate their own LMR networks, separate and distinct from those used by police, fire, and EMS.

- The United States population continues to increase, and continues to further concentrate in urban, densely populated, regions. This fact compounds the effects of disasters (earthquakes, hurricanes, tornadoes). Furthermore, terrorism is a real concern in many cities. These challenges have spawned emergency operations centers and functions to coordinate agencies' response to major incidents and disasters. The need for communications interoperability between jurisdictions and agencies has never been greater.

- Public safety operations in the United States are highly decentralized. There are more than 18,000 law enforcement agencies[5] ranging in size from two officers to the nearly 36,000 officers in New York City.[6] There are more than 22,000 fire departments,[7] most (but not all) of which also handle emergency medical services. There are hundreds of thousands of villages, towns, cities, counties, and special purpose districts responsible for public safety in their geographies.

- Coordinating the communications needs for all these diverse agencies and jurisdictions is difficult. Most areas of the country have long-standing turf battles over geography and budgets, which has led to a balkanization of how LMR networks are constructed and operated.

All of these factors have spawned a wild diversity of voice land-mobile radio systems in the United States, operating in various different parts of spectrum, differentially funded, in various states of age and obsolescence, and, for the most part, not interoperable. To be blunt, a $5,000 handheld radio that works just fine in one jurisdiction is a $5,000 hunk of useless plastic and circuitry in most of the country.

Luckily, the Middle Class Tax Relief and Jobs Act of 2012, if properly implemented, will create an interoperable nationwide data[8] network serving most public safety operations in the United States. If improperly implemented, it will result in just adding more network diversity and confusion to the existing situation.

Innovative approaches to building and operating the nationwide network

Seven billion dollars is a lot of money. But building a nationwide 4th generation wireless network could cost $30 billion to $40 billion or more. Such a network would require 40,000 cell tower sites or more. And operating it will cost billions more each year, a cost that must be paid by the users of the network.

Furthermore, no public safety agency will *have* to use the network! Indeed, the law specifically allows states to opt out of the plan. Many cities and counties have love-hate relationships with each other and with their states. And all state and local jurisdictions have that love-hate relationship with the federal government, which is charged to build this network.

In addition, the technology used in building this network—long-term-evolution or LTE—is, well, *evolving*. LTE is an international standard. The current "version" is Release 8.[9] Hundreds of commercial telecommunications carriers will be using LTE technologies to build their networks worldwide, but it is still relatively new and rapidly changing. LTE supports data networking and Internet access, but it does not yet support wireless voice phone calls. It will be seven to 10 years or more before LTE supports the voice dispatching and direct radio-to-radio communications required by public safety agencies.

Figure 2. The LTE Logo, trademarked by the 3GPP Standards Organization.

These are just a few of the challenges faced by the FirstNet authority as it builds the Nationwide Public Safety wireless Broadband Network (NPSBN).

Bold, innovative approaches are needed to overcome these challenges, plus all the problems, some of which were described above, associated with traditional LMR networks. Here are a few such innovative approaches that could be used to construct the NPSBN:

Testbeds and pilots

Building the NPSBN is a gigantic technology project. Given the relative immaturity of the LTE technology and the vastness ("nationwide") of the work, pilot projects or testbeds would allow FirstNet to test different project management and technological approaches.

But given the urgency of the need for the network, is there time for pilots?

Luckily, the NPSBN comes complete with a set of pilot implementations already in progress.[10] In May 2010, the Federal Communications Commission (FCC) granted "waivers" of its rules to 21 states, cities, and regions. These waivers allow those jurisdictions to build LTE networks to serve their public safety needs. Furthermore, NTIA the same agency that will be creating the FirstNet Authority—allocated more than $300 million in grants[11] to seven of those jurisdictions to help them build the network. Some of the remaining 14 waiver recipients are using other grants, their own state and local funds, or other innovative approaches to build networks in their area.

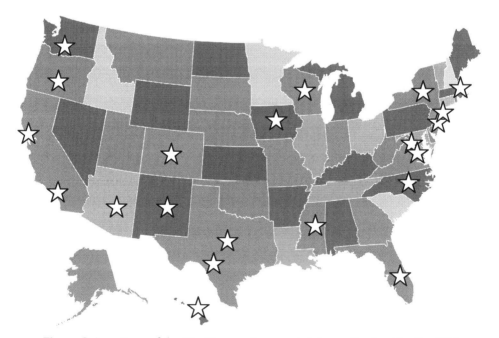

Figure 3. Locations of the 21 cities, regions, and states authorized by the FCC in May 2010 to build public safety data networks.

When the spectrum license is assigned to FirstNet, it could immediately terminate the authority of all these waiver recipients to build and operate their networks. But FirstNet also could embrace these projects-in-progress, and even encourage further pilots and experimentation by some or all of the 21 waiver recipients. In so doing, it not only tests a variety of approaches for the larger network construction projects, but it also leverages hundreds of millions of dollars in other federal, state, and local funds.

Top-down or bottom-up?

In the past, faced with a project of this size, the federal government typically would turn to one or more of several dozen private contractors to manage the project.[12] And the Spectrum Act specifically directs the FCC to create interoperability requirements[13] and FirstNet to use those requirements to issue RFPs for network construction.

There are, however, alternative approaches to this "top-down" approach. Here are examples:

- FirstNet could issue RFPs and award individual contracts to several different private contractors that would be system integrators for actual network construction. Later, FirstNet could issue RFPs for entities to operate the network. FirstNet might then allow states, and perhaps even local governments, to build the network in their areas to a set of nationwide architectures and standards, using those contracts.

- FirstNet could issue RFPs and award individual contracts for a wide variety of goods and services used in the network construction, e.g., central network equipment, user devices, cell sites and towers, network management systems, and so forth. Then, it might allow states, and perhaps even local governments, the authority to manage network construction in their areas to the nationwide architecture/standards using those contracts.

- FirstNet might allow state and local governments to bid on RFPs to manage the network construction in their area.

- FirstNet might allow state and local governments to separately bid on contracts to operate the network in their area once it is constructed.

All of those approaches envision a much stronger leadership and partnership role for state and local governments in network construction and operations. States and local governments know the "lay of the land" in their areas, including potential users, assets, and problems that need to be addressed. By building and operating the network in their geographies, those governments have "skin in the game" and are encouraged to use their own assets in construction. Furthermore, this approach encourages states/locals to include a wide variety of users—beyond just first responders—spreading costs across a wider user base.

Expansive definition of public safety

Many traditional LMR networks are limited to law enforcement, firefighting, and emergency medical services. These networks have a relatively narrow user base, perhaps only 4 million first responders nationwide. This narrow user base significantly increases the per-user cost of network construction and operation.

FirstNet must include a much more expansive view of the user base, and this broader definition is allowed by the Spectrum Act. At the very least, any local, state, or federal agency involved in a public safety activity should be on the network. This would include electric, gas and water utilities, public works, transportation and roads departments, transit, public health and hospitals, even building and development departments, and animal control.

Indeed, network use could be expanded to include any such function (e.g., utilities) whether publicly or privately owned, plus general government functions and also fixed uses, such as meter reading, smart grid, and video cameras. Imagine, for example, using the network to read, manage, and control 400,000 electric meters in a smart grid.[14] Control means not just reading the meters, say every 15 minutes, but also allowing the electric utility to shut them off, turn them on, manage and determine outages, and so forth. At first thought, this use seems a waste of precious bandwidth. But actually, most of the network will have plenty of capacity most of the time. Meter management requires trivial bandwidth—a few bytes of data every few minutes for each meter.[15] Using the many priority-setting capabilities of LTE, such reading can be placed at the very lowest priority of network use, and even shut down when the network is overloaded. And such use brings enormous value: say the electric utility paid 50 cents per meter per month for this service—that's $200,000 in operating income a month, $2.4 million a year to operate the network. And such use allows individual homeowners and the utility to see and manage their electric use in a much more granular fashion than traditional once-a-month reads.

Other potential uses, however, are not so trivial. Fixed video cameras could have a huge effect on network bandwidth. A single high-definition television camera might stream 6 million bits per second onto the network. These intense uses will mean the video cameras really should be connected with fiber optic cable operating at gigabits (billions of bits) per second. This reserves the NPSBN bandwidth just for sending video to mobile responders, not providing backhaul for the individual video cameras. But, again, LTE networks provide many methods to minimize the impacts of such video, e.g., the prioritization schemes or transmitting very low frame rates (one frame every second or less) until such time as an incident occurs and higher quality video is required.

Public-public partnerships

In building the NPSBN, FirstNet must leverage assets already put in place by cities, counties, states, and other local jurisdictions. Such assets could include existing radio sites, other public buildings and sites, data centers, fiber optic and microwave networks, and other assets. These local and state jurisdictions have already purchased and invested in such assets. This use of existing assets has several positive benefits:

- Such use reduces the capital expense involved in constructing the network by using existing in-the-ground assets owned by states and locals, such as radio sites and fiber optic networks.

- Quite often, these assets are already hardened to public safety standards. "Hardening" means the asset is prepared to withstand almost any conceivable disaster. For example, a radio site may already have a backup generator, a tower may be built to withstand hurricanes and earthquakes, and a site may be connected by fiber optic cables running two or three different directions away from the building so if one cable is cut, the other connections still operate.

- Such use of existing assets reduces the ongoing expense of operating the network because, for the most part, state and local governments maintain and operate these assets for their other mission-critical functions anyway.

- By using such assets, state and local governments will be more likely to actually become users of the network, as they have "skin in the game."

There are also challenges in these public-public partnerships between the federally established FirstNet and state and local responder agencies. Examples of such challenges:

- How do we value these assets and compensate jurisdictions for their use? Perhaps a city brings an entire loop-protected fiber optic network to FirstNet.16 Would FirstNet pay market value for such a network (and what is the "market value")? Or would the jurisdiction's agencies receive a 10% (or 30% or 50%) reduction in its monthly subscriber fees? Should FirstNet, when obtaining such assets, do so by RFP, where local governments would bid against commercial providers?

- How would FirstNet legally obtain and own such assets? Should local governments continue to own the asset but provide them to FirstNet on long-term leases or contracts? Does FirstNet have to "own" all such assets, including electronics such as the eNodeB equipment at cell sites? What happens when one party wants to terminate the relationship or replace the asset mid-contract?

- Are the assets really public safety grade? Are they well-maintained and regularly upgraded? How would a local government keep the upgrade of such assets "in sync" with upgrade of similar assets elsewhere on NPSBN (e.g., router, switch, and server operating systems)? Are backup generators, uninterruptible power supplies, and similar hardening in place and well-maintained?

- Would local assets be secured from both physical and cyber threats? Do we need best practices and checklists and regular inspections to ensure security? Do we need firewalls between locally maintained assets/radio access networks and the wider NPSBN?

- How much flexibility will state and local responders have in selecting user components, whether user devices or two-factor authentication schemes or applications? More flexibility for a local agency implies potentially lower costs, better adaption of the use of the NPSBN to local conditions, and improved adoption. On the other hand, such flexibility could inadvertently introduce network performance issues or security threats.

There is tremendous opportunity for FirstNet to actively engage state and local governments in all phases of the NPSBN planning and construction. Earlier and closer involvement will probably result in wider acceptance and use of the NPSBN by local and state agencies. But, with 50 states, hundreds of larger local governments, and a number of tribal nations involved, "close" involvement could be a herculean task for FirstNet.

Private-public partnerships

FirstNet clearly will leverage considerable use of privately held assets, either through partnerships or RFPs to obtain equipment and services. Such assets could include:

- Cell towers and sites. AT&T, T-Mobile, Sprint, Verizon, U.S. Cellular and other telecommunications companies, plus other providers such as American Tower or Crown Castle, own hundreds of thousands of cell sites nationwide. Many such sites have spare capacity on the tower or in the building, which FirstNet might use;

- Backhaul and interconnection services, whether local, regional, or nationwide (e.g., fiber and copper networks, IPX services, data centers);

- Hosted services, such as evolved packet cores[17] and related services or eNodeB[18] devices;

- Applications, applications stores, and application development. Commercial apps stores such as iTunes or the Android Marketplace contain hundreds of thousands of such apps. First and second responder apps stores will undoubtedly develop as the NPSBN is built; and

- User and subscriber equipment, whether purchased or leased. Such equipment might be smart phones, tablet computers, video cameras, sensors, and many other pieces of equipment.

For the most part, private telecommunications companies and other private companies are well acquainted with how to procure and deploy such services. Partnerships between these companies and FirstNet should speed the NPSBN deployment.

Devices and user equipment

One of the most persistent complaints about LMR equipment is the cost. A single public safety radio might cost $5,000 or more. Many public safety departments, when seeking to acquire such equipment via RFP, add a wide variety of requirements from ruggedness or waterproofing to button size (large buttons required for use with gloves) or long battery life.

Agencies need to examine the need for such requirements. Rather than having a single subscriber unit ("handheld radio") which is rugged, hardened, and has a large battery for long life, it is probably easier to use commercially available equipment. Instead of having one rugged device costing $1,000, $2,000 or more, public safety officers and other officials could be equipped with a commercial smart phone and tablet at much less cost. Agencies could even provision responders with two smart phone devices (i.e., a spare in case of damage) and a tablet less expensively than a single ruggedized device.

Another interesting concept for the NPSBN is ubiquity. Should every consumer and commercially used mobile phone in the United States have 700 MHz public safety capabilities? This potentially allows transmission and reception of public safety data from every mobile phone. Of course, such phones will use commercial services from telecommunications carriers for all normal voice calls and other functions. But during incidents or emergencies, public safety agencies might use the 700 MHz network to send an alert to all mobile phones in a specific geography or location. Such alerts might be a warning about a traffic jam, SWAT team action, or other incident. Or the alert could be a request for information, such as photographs, tweets, or video about a developing incident or problem. In such a fashion, public safety departments could "crowdsource" information to manage the disaster or other incident.

This idea has issues with cybersecurity and hacking, which might imperil the usability of the NPSBN. And other emerging alerting technologies (such as CMAS[19]) might be less dangerous to implement.

Summary

The Nationwide Public Safety wireless Broadband Network is an opportunity for the United States to "get it right" in creating a truly interoperable nationwide network for use by all responders and government agencies at all levels. But, construction of the network will require innovative approaches and a true collaborative partnership between the First Responder Network Authority, states, local governments, and private companies.

BILL SCHRIER, *served the city of Seattle for nearly three decades, with the last third of that as head of IT and CTO/CIO. Having retired from the city, but not from the IT arena, Bill is now the Deputy Director of the Center for Digital Government, owned and operated by e.Republic. Bill was elected as a PTI Fellow in 2010 and was also elected as a Fellow of the Association of Public Safety Communication Officers. Schrier has been very active over the past few years with public safety communication systems, and in particular the designated national "D" Block. In his earlier career, Bill was a police officer in the City of Dubuque, Iowa.*

ENDNOTES

1. All wireless devices—broadcast television, AM/FM radios, computers with Wi-Fi, cell phones, smart phones, even garage door openers—use radio spectrum. The Federal Communications Commission manages and oversees how spectrum is allocated and used in the United States. With the explosion of use and numbers of cell phones, smart phones, tablet and laptop computers, the demand for such spectrum by wireless telecommunications carriers and others is skyrocketing. In the Spectrum Act, Congress directed the FCC to reallocate certain parts of spectrum from less intensive uses and auction it to the highest bidder. Such auctions might raise $24 billion to $26 billion over the next 10 years.

2. "NPSBN" is the usual acronym used for the network and, therefore, it is used here. That acronym is technically incorrect, however. This is a *wireless* network, as distinguished from a fiber optic broadband network or a wired network, which are vital to the Internet and all forms of public safety and general government communications.

3. Congress had set aside 10 MHz for the use of public safety. The Spectrum Act takes this 10 MHz and adds an additional 10 MHz, often called the "D Block" to create the 20 MHz to be used by NPSBN. The term "D Block" is sometimes erroneously used to describe the entire NPSBN effort.

4. The 700 MHz band is very valuable because these radio waves will penetrate buildings to a certain extent. This is very useful for firefighters, emergency medical technicians, and police officers who often must respond to incidents inside buildings. Verizon, AT&T, and certain other cellular telecommunications companies also have licensed spectrum in the 700 MHz band for use in their consumer and commercial networks.

5. http://bjs.ojp.usdoj.gov/index.cfm?ty=tp&tid=71

6. http://en.wikipedia.org/wiki/New_York_City_Police_Department

7. http://www.usfa.fema.gov/downloads/pdf/publications/fa-278.pdf

8. The NPSBN is a data network and will not replace LMR voice networks at this time. Many hope the technology will advance and eventually allow the NPSBN to handle voice dispatch, but significant technology and standards hurdles stand in the way of that goal.

9. http://www.3gpp.org/Release-8

10. Indeed, the City of Charlotte and Harris County, Texas, are both slated to make their networks operational in summer 2012.

11. The grants were under the Broadband Technology Opportunities Program, or BTOP. The availability of the grants was announced on the same day as the FCC announced the waivers—May 19, 2010. The grant awardees were announced in September 2010. The grants went to the City of Charlotte (NC), the states of Mississippi, New Jersey and New Mexico, Adams County (CO), the Los Angeles Regional Interoperable Communications System (LA-RICS), and to Motorola to construct a network for the San Francisco Bay Area.

12. Examples of such contactors include Lockheed Martin, Northrup Grumman, General Dynamics, IBM, Scientific Applications International Corporation (SAIC), Computer Sciences Corporation (CSC), Raytheon, and others.

13. The Interoperability Requirements are almost complete. See http://urgentcomm.com/interoperability/news/broadband-interoperability-report-20120524/

14. This example is based upon the City of Seattle's LTE network design, which called for about 40 cell sites and might have 5,000 "traditional" mobile data users/units in a city workforce of 11,000 employees. The city also has a municipal electric utility, Seattle City Light, with 400,000 electric meters, which could be networked into a smart grid.

15. The data capacity of a single LTE cell site is 100 million bits per second or more.

16. Seattle has about 1,600 miles of streets and has government-owned fiber optic cables on 500 of those miles, reaching every neighborhood, every school, every library, every fire station, and most other public buildings within the city limits.

17. The Evolved Packet Core (EPC) is the central "switch" of an LTE network. A half-dozen EPCs can successfully route data and voice for 100 million users or more.

18. eNodeB devices are the equipment at a cell site which allow transmission and receiving of LTE signals at that site.

19. CMAS: http://transition.fcc.gov/pshs/services/cmas.html

Chapter 12: Sensory Overload Considerations for Next Generation 9-1-1 Systems

BY RONALD P. TIMMONS

Introduction

As agencies deploy Next Generation 9-1-1 (NG911) systems with video, images, and text messages, consideration must be given to human factors design for Public Safety Answering Points (PSAPs). Additional video displays will compete for the attention of PSAP staffs typically responsible for data on more than two video displays per workstations. It is not yet clear how many displays are going to be attempted at typical NG9-1-1 deployments, but with screen space already crowded with existing applications, more screens and consolidated applications seem likely.

Due to complex limitations of the human condition, there is a fine line between too much information and not enough. Failure in the PSAP environment can result in dire consequences for crews and citizens awaiting emergency service. Emergency communications personnel do not have the luxury of deliberation and debate. Understanding the decision-making environment of the PSAP is necessary before designing systems to meet the unique circumstances experienced by PSAP staffs. This chapter provides technology leaders with an overview of the situation and recommendations for optimal ergonomic configurations.

The 9/11 Commission found that there is fallacy in the tendency to expect operational improvement merely by the deployment of new technology (National Commission on Terrorist Attacks 2004, 280-285; U.S. Fire Administration 1993, 3, 7). The Commission also articulated an expectation for PSAP management to maintain detached overall perspective when major incidents are under way, with personnel poised to deliver sound advice to additional callers. While a new system may provide opportunity to gain additional information and to achieve better situational awareness, there is also a possibility of sensory overload and "too much information."

It is easy to assume that if more information is supplied to PSAP personnel, they will be better informed and more effective. In workplaces as diverse as commercial airline cockpits, nuclear power plant control points, military weapons war rooms, and security guard control desks, people use information displays with the expectation of enhanced ability to do their jobs. Each of those places has been the site of failures and missteps leading to catastrophic outcomes. The security industry has used surveillance cameras for years as an adjunct to actual physical presence. The expectation of cameras being a force multiplier has had mixed results. Thefts have been discovered while in progress, but it is often the review of archived data that proves most useful after an event.

Decision making in the PSAP environment

While it is natural to focus on the technological aspects of making new technology work end-to-end, the overarching purpose for NG9-1-1 should be decision support for PSAP personnel. The interaction of people and events generate geometric complexity in the decision-making process used by communications personnel. There are significant, complex mental processes occurring in the milliseconds between input of conditions stimuli and the resultant selection of an appropriate response (Gladwell 2005, 11-15). CIOs need to gain awareness and appreciation for the way people use information in high-pressure PSAP situations.

The need to establish "situational awareness" and a "big picture view" are terms used to acknowledge the tendency of those at the communications center level to take a myopic view of circumstances immediately apparent, to the exclusion of considering the overall situation. Such "tunnel vision" is typical as people struggle to answer the concern of the moment. Development of situational awareness is a necessary ingredient in sound crisis decision making by PSAP personnel, and effective processing of information received is an important part of gaining the necessary situational awareness.

Paradoxically, those involved in the dynamic environment of emergency services traditionally have a systemic resistance to change. Personnel in public safety agencies tend to function in rigid, linear organizations in which inflexible rules convey organizational expectations and standardize the approach to situations encountered. Rapid, adaptive learning in "real-time" has to occur, yet those involved have classic attributes of traditional, rules-bound bureaucrats. The same hierarchical orientation from daily para-militaristic practices requires reversal by the communications staff, as they seek creative, flexible decision solutions in intense communication situations.

It is important for CIOs to consider the multiple levels of activity and the impact of expected and unexpected events on people operating in high-reliability situations. Subsequent behaviors redirected by unexpected events make analysis of PSAP decision making very complex.

Military parallel

As PSAP personnel struggle to develop greater comprehensive awareness of events occurring in the community through a variety of technological means, other venues have similarly frustrating history of trying to improve the human-machine interface. Parallels in military culture exist, as similar situations have led the armed forces to pursue adaptive information processing behaviors.

In the U.S. Navy, "having the bubble" is the phrase for maintaining a big picture view of operations (Roberts and Rousseau 1989, 135; Bigley and Roberts 2001, 1292). Much like counterparts in the military, our community emergency communications personnel face overwhelming variables, chaotic situations, critical decisions, and death of residents and coworkers as a potential consequence of poor decisions.

Vincennes

The U.S. Navy has undertaken significant efforts to better understand how their personnel react under stressful communications situations. The 1988 downing of a commercial airliner by the USS Vincennes offers insight into the challenges of making critical time-compressed decisions, based upon monitoring of decision support system displays. Personnel made fatal errors despite the availability of significant technological assets, which represented the state of the art at that time. In one of the most studied events involving people using technology to glean information and make decisions, a number of vulnerabilities and shortcomings were identified.

After the Vincennes incident, the Navy searched for greater insight into how people react in high-stress environments and how to improve the quality of decisions by recognizing the human factors aspects of the technology provided. The Tactical Decision Making Under Stress (TADMUS) program was initiated in 1990, with the goal of designing a decision support system (DSS) based on understanding the cognitive strategies people use when making tactical decisions. The human-computer interface design principles were explored to compensate for human cognitive processing limitations (Hutchins 1997, 207-215).

The TADMUS studies found that the sheer volume of information flowing in a compressed timeframe produces a high cognitive workload environment. Short-term memory degrades under stress, further limiting the cognitive capabilities of the individual. Performance was improved when decision support, in the form of a user-centered DSS was provided to the decision makers. Situational awareness was enhanced by improving the visual displays provided. The DSS had modules to help keep track of actions taken or awaiting attention, as opposed to the traditional expectation of the individual relying upon memory to remember which tasks need to be performed and when. Other modules synthesized parametric data and provided graphic presentations to depict what tracked aircraft were doing along with recommended actions to be taken.

Decision makers are susceptible to cognitive biases when operating under stress, i.e., high workload, time pressure, and information ambiguity. People perform best when provided with visual prompts containing prioritized information in simplistic syntax. Decision makers also are susceptible to cognitive "tunnel vision," a situation where they focus on a narrow set of cues when under stress, even to the exclusion of critical matters awaiting attention. Another example of commonality between military and PSAP decision making is the condition of decision confirmation bias: commanders tend to build upon assumptions made earlier in the decision sequence, even if they were based upon limited or even erroneous assumptions.

Physiological and emotional components

There are many and varied reasons why people do not make the optimal use of information provided to them. Physiological influences detrimental to cogent information processing and articulation of optimal communications are a major factor in the quality of decisions made by PSAP staffs.

Emergency communications personnel experience a state of "expressive suppression," which is the conscious inhibition of emotional expressions while emotionally aroused. PSAP staff force themselves to "stay calm" and to control their emotions as they face the formidable challenges present in handling emergency calls. People who suppress natural emotional responses experience elevated blood pressure, increased stress levels, disrupted communications, a reduction in rapport building, and inhibited relationship formation (Butler et al. 2003, 48-67).

- Emotions play a significant role in emergency operations, and communications personnel remote from the incident are similarly impacted.

- Dispatch staff cannot isolate themselves entirely from the effects of emotion, despite expectations of peak efficiency.

- Emotions are triggered by the dynamic and intense conditions of emergency situations; it is not possible to separate emotions from the decision-making process.

Visual and audible data from NG9-1-1 will have to be appropriately detailed, yet measured and managed to acknowledge the compromised state of PSAP personnel during intense operations. PSAP personnel have diminished capability to make sense of data during intense operations, right at the very time it may be needed most.

How communications personnel make decisions

People in everyday life form impressions and make decisions based on observation of nuanced vagaries. "Thin slicing" is one way to discern these factors. Gladwell (2005, 23) defined thin slicing as "the ability of our unconscious to find patterns in situations and behavior based on very narrow slices of experience." Discovery of new pieces of information allow communications personnel to make sense of dynamic variables, thus allowing the situation to become linear, predictable, and controllable.

There are several steps in decision making; they essentially are:

- gather information,

- process it, and

- apply the best choice.

After scrutinizing available data and alternatives, choices are applied in a rapid, routine, and reflexive manner by PSAP personnel. Emergency communications personnel attempt to apply known information in milliseconds. Contrast that with corporate managers who can sometimes take weeks to reach decisions with far lower consequences. PSAP staffs work in high-stakes environments where there is little time for debate, nor contemplation. Testing of counterfactual alternate endings is not possible in the emergency environment. Emergency personnel tend to use recognition-primed decisions, employing the first workable solution in the interest of time and expediency (Klein 1999, 30).

Design of systems intended to support those facing complex problems should acknowledge the limits of cognitive capacities and narrow analysis to the most critical, non-exhaustive components. Any attempt to take every possible factor into consideration leads to the likely outcome of narrow definitions and a resultant limited scope (Lindblom 1959, 80-83).

Former Secretary of State and retired U.S. Army General Colin Powell waited for 40 percent certainty on data before considering an intuitive solution to a problem. To overcome what some may call "analysis paralysis," Powell viewed any more than 70 percent certainty to be more information than necessary to make a sound decision, wasting valuable resources on data acquisition (Klein 2003, 171).

NG9-1-1 displays should have key information readily available, but attempts to display everything known will similarly waste precious cognitive processing and information of lesser importance.

Framing and heuristics

The term "decision framing" refers to the situation whereby acts, outcomes, and contingencies attributed to each choice in a decision sequence shape the decision maker's perceptions. Framing is influenced by both the circumstances prevailing within the decision and the habits, customs, and norms possessed by that particular decision maker (Tversky and Kahneman 1981, 453). Therefore, there are two variables present in any major decision: the person and the circumstances.

People make decisions from their own neutral reference point, making the location of the reference point, i.e., positively or negatively framed, an influence on the amount of risk assumed (Bazerman 2006, 43). The decision maker may select a different course of action, if there was awareness of a slightly different decision frame or ease in which to access and consider other alternatives. Decision frame also gives the decision maker a comfort zone of operation, in which the single option chosen early in the process becomes more and more fortified, shaping the point of reference and increasing risk propensity (McDermott et al. 2002, 145).

Herbert Simon's bounded rationality theory holds that actors in intensive decision-making situations use a form of "rational choice that takes into account the cognitive limitations of the decision maker—limitations of both knowledge and computational capacity" (Simon 1997, 291). There is a tendency to apply heuristic decision making as reflexive coping mechanisms. Heuristics are short cuts or "rules of thumb" used to transfer familiar experiences into new situations with presumed similarities.

PSAP personnel bring heuristic orientations with them to the workplace and rely heavily upon them to seek positive outcomes. Use of heuristics occurs especially in situations of uncertainty, where available options are imprecisely specified. People prefer heuristics under unfamiliar or critical circumstances because they are cerebrally economical, demanding fewer cognitive resources to reconcile the dilemma presenting itself.

The use of "gut feelings" sometimes requires an umpire to assure that the logical mind has an opportunity to seek any disproving evidence through the objective examination of fact. A gut feeling can serve as a re-opener when something does not seem quite right about a pending decision; gut feelings are "unprovable" and an admittedly unquantifiable phenomenon. Data from NG9-1-1 will play a role in providing both confirming or disproving information, providing that the data can be quickly and accurately deciphered.

Tversky and Kahneman (1974, 1124-1131) called attention to three areas of heuristic and cognitive bias concerns:

1. Representativeness, in which one facet of the decision is influenced by the other. In this filtering situation, preceding actions taint the final decision. Another way of looking at representativeness heuristic is gambler's dilemma, in which the gambler expects a certain outcome to be "due" despite equal odds of any of the possible outcomes occurring next.

2. Instance and scenario availability, during which people recall personal experience with a circumstance, and disproportionately assume recurrence of factors despite nuanced differences and often-disparate time and conditions from the base experience triggering the recollection.

3. Adjustment from an anchor refers to situations in which bias influences people to pick a starting point presuming an outcome, and/or to tend to stay close to an expected outcome and rearrange the circumstances to fulfill preconceived expectations.

Why critical information is missed

Distractions

Once NG9-1-1 videos, images, and text are flowing to PSAPs, there will be an expectation that personnel will be able to use it all equally and effectively. Sports fans will readily recognize the phenomena in which a referee makes a judgment call, only to have it look differently upon video replays and from other angles. People miss obvious information all of the time. How can that be?

Perception of a change occurs most effectively when the object is given focused attention (Rensink et al. 1997, 372). When there is a lot of changing stimuli observed, it is unreasonable to expect all unusual events to receive equal attention. Distraction conflict theory investigates the influence of distractions on decision performance.

Speier et al. (2003, 773-774) used distraction conflict theory as the basis for building their interruption/decision-making model. Their findings included:

1. Capacity interference results when the amount of incoming cues is greater than the processing ability of the decision maker.

2. Interference occurs when the decision maker must attend to two or more inputs requiring the same physiological mechanisms.

3. As people attend to multiple demands for attention, they may be unable to access the necessary amount of additional cognitive processing capabilities needed for successful problem resolution.

4. Memory loss and confusion are likely byproducts of such circumstances.

While processing critical information, people tend to minimize their expenditure of scarce cognitive resources, which results in a lack of scrutiny of both relevant and irrelevant new information (Speier et al. 2003, 775). New technologies do not guarantee immediate awareness enhancement; instead, they may actually complicate matters, as we have seen in the recent distraction debate about driving while talking on cell phones or text messaging.

Interruptions in train of thought are a regular feature of the PSAP environment. One study showed that managers spend 10 minutes out of every hour at work responding to interruptions (O'Conaill and Frolich 1995, 262-263). In 41 percent of the cases studied, people do not return to the original task after interruption. It is not just the actual time of the interruption to consider. The decision maker must cognitively re-immerse in the train of thought prevailing before the interruption during a recovery period. Communications personnel experience a constant stream of interruptions and redirection of their attention through contacts with people over two-way radios, telephones, and in person.

Dispatch personnel face another challenge in their time-sensitive environment. People in all fields of endeavors find value in taking a break when faced with a situation in which they are stuck in finding the solution to an issue at hand. After relaxing or attending to other matters and then returning to the work, the solution often spontaneously emerges into consciousness. Taking a break and "quiet time" are sometimes recommended as antidotes to debilitating information overload and constant interruption. Such "time-outs" are valuable to our decision-making abilities, yet time is a luxury not usually available to PSAP professionals. Those recommendations may work in some venues, but PSAP personnel deal with time-critical matters, with little to no time for fact finding and debate.

Like many aspects of decision theory, the absolutes elude us when it comes to interruptions. One team of researchers (Speier et al.1999, 340-348) tested the notion of interruption as a welcomed and valued element in the decision-making process. After observing subjects as they perform in a problem-solving environment, it was found that interruptions used to gather more information input and time for collaboration helped to improve the quality of decisions made on "simple tasks." Task complexity is increased by making later decisions dependent on imbedded decisions made earlier in the sequence. The quality of decisions degrades upon interruption during complex tasks, yet the quality improves after interruption when working on simplistic, monothematic issues. Carefully designed improvements in the human-machine interface hold potential for improved decision making in critical situations, as communications personnel use an increasing array of technologically complex communications and information management equipment.

Inattentional blindness

When significant information is missed on a video display or audio source, the inevitable question is how could they have missed it? Bazerman (2006, 171-172) attributed the association of traffic accidents and cell phone use to be a result of inattentional blindness, a phenomenon in which the majority of those viewing extreme features of an event do not perceive critical elements. For instance, spouses frequently accuse the other of not retaining something just told to them by their partner.

In a well-known study written by Simons and Chabris (1999, 1066-1073), test subjects were instructed to focus their attention on people playing a basketball bouncing game. The video features the introduction of a gorilla into the video frame, or more exactly a person in a gorilla suit, who walked across the screen in front of basketball players. The person in the gorilla suit did not just flash across the screen but rather stopped in the middle of the players, as the action continued around the gorilla-suited person. The person in the gorilla suit actually turned and faced the camera, thumped his chest, and resumed walking at a leisurely pace across the field of vision. Half of the people who observed this demonstration did not see the person in the gorilla suit at all because they focused intently on counting the number of the passes between members of the different basketball teams.

The "Gorilla" experiment was an update of an earlier study in which a woman with an umbrella walked across the test subjects' screen. The study, conducted by Becklen and Cervone (1983, 601-608), showed a similar inattention bias to those who observed the incident because they focused their attention so sharply on the area of interest. In the Simons and Chabris study, one team of the three basketball players wore all white; the other team of three wore all black. Since the gorilla suit was all black, they were able to test and control for the influences the colors and inattention had on whether subjects noticed the gorilla walking across the frame.

In a counter-intuitive result, the study demonstrated that those who were studying the black-suited players were more likely to notice the similarly colored black-suited gorilla walk across the screen, instead of those who were concentrating on the players dressed in all white. This differed from the expected outcome since we may anticipate they would notice something contrasting from the white-suited players. However, the researchers concluded the observers focused so intently on those similar to the object of their attention that they were not able to see the opposite effect.

The "Gorilla" study "required observers to attend to one event *while ignoring another* that was happening in the same region of space" (Simons and Chabris 1999, 1072). This is a phenomenon of "directed ignoring," which inhibits perception of not just the event the individual is attempting to ignore, but of all unintended events occurring in the space. The subjects involved in this experiment were more likely to notice an unexpected event that shares the same basic visual features, which in this case was color. This would be the opposite of the phenomenon expected in which an item that differs in basic features from the rest of the display might be easier to notice and identify. With approximately half of the observers failing to notice the highly contrasting, but unexpected, event on the video screen, there was "a robust phenomenon of sustained inattentional blindness for dynamic events" (Simons and Chabris 1999, 1069).

Similar inattention is noted when individuals focus intensely and exclusively on relatively static video displays, only to miss a crime or significant issue on the screen. This can explain why security camera views often become routine and monotonous, and therefore, critical events go unnoticed. Understanding of the complex human factors involved beyond the capabilities of even the most robust hardware imaginable is necessary to design video displays to optimal advantage.

Multitasking myth

PSAP personnel pride themselves on an elusive ability to "multitask." It is likely that they are rapidly moving from one task to another, in seemingly seamless fashion, yet the ability for people to fully perform more than one critical task simultaneously is dubious. The performance of people degrades as multiple demands are imposed simultaneously, amounting to a condition of sensory overload. Information overload is the expected result when the volume of input exceeds the processing capability of those involved (Milord and Perry 1977, 131-136; Speier et al. 1999, 338).

Emergency personnel experience sensory overload, leading to ineffective communications.

- Sensory overload plays a role in the non-technical impediments limiting the effectiveness of PSAP staff members.

- Distraction is typical by the multiple conversations, noisy environments, and multi-sensory demands for their attention.

- Even when successful delivery of a message occurs, the recipient frequently misinterprets it or lacks awareness that the message occurred (Timmons 2006, 2007).

- Emergency personnel operating in critical decision-making situations experience sensory overload leading to communication deficits.

- Rapid movement from one critical event to the next inhibits cognitive recovery before the next intense matter with differing dynamics is addressed.

- Despite intentions to the contrary, there are significant impediments in the multi-sensory world of PSAP video displays when attempts are made to provide exhaustive information.

Physiological deficits

In discussing the effects on the human body in stressful situations, Gladwell entitled a subchapter in his book *Blink* (2005, 221-229) "Arguing With a Dog," based on the experiences of a retired army lieutenant colonel and his observations of people functioning in a state of extreme arousal. As heart rate increases in response to threatening or stressful circumstances, performance actually improves in the 115 to 145 beats per minute range. After 145 beats per minute, motor skills become difficult. After 175 beats per minute, cognitive processing begins a selective shutdown sequence with brain functioning then becoming similar to that of a dog.

Police officers involved in shootings often experience altered states of sensory cognition. They report, "Extreme usual clarity, tunnel vision, diminished sound, and the sense that time is slowing down. Our mind, faced with a life-threatening situation, drastically limits the range and amount of information that we have to deal with. Sound and memory and broader social understanding are sacrificed in favor of heightened awareness of the threat directly in front of us" (Gladwell 2005, 224). Those impacted do not even have to be present at the incident scene. Dispatch personnel undergo physiological stress reactions in response to what they are hearing from callers and first responders.

Solutions

Teamwork

There is a limit to the amount of sensory input PSAP personnel can manage before they are overwhelmed, to the detriment of effective operations. One aspect of managing the sensory overload experienced by PSAP staffs involves greater teamwork and additional staffing to parse critical information and provide redundant safeguards.

Gladwell (2008, 184-185) found that a disproportionate number of aircraft flights with tragic outcomes had flight crews not used to working together. The optimal flight deck approach is when partners catch each other's errors, rather than one trying to do too much and the other there ostensibly just to take over in the event of complete incapacitation of the other. Any teamwork in the dispatch center needs to feature rapport, boundaries, expectations, and anticipation of each other's tendencies.

There is a fine line between just enough teamwork and too much. Additional involvement of adjacent dispatch coworkers may exacerbate the distraction level influencing PSAP staff and their ability to maintain clear and immediate focus on radio and telephone conversations.

Human factors engineering

Ready access to information has the risk of debilitating instead of facilitating. Properly designed equipment provides greater situational awareness and better quality of decisions reached using the information from the technology. Klein (1999, 279) noted that in the past, "information was missing because no one collected it; in the future, information will be missing because no one can find it." Information overload in the emergency communications environment makes it impractical to bombard emergency communication staffs with more and redundant information.

The challenge posed to the human factors community is to devise technological improvement in the human-machine interface to allow critical information to flow without overwhelming the PSAP staff with sensory overload. The auto industry has been one of the leaders in human factors research with similar concerns on how much information and optimal syntax to display on vehicle video monitors.

CIOs need to consider systems design provisions and workstation ergonomics to assure information is accessible when needed, but not superfluously squeezed into the field of vision and attention of people performing in the multisensory PSAP environment.

Data displays

Appropriate human factors engineering to produce supportive information displays, and user-centric design to avoid information overload, are necessary to guard against overly optimistic expectations. When data is displayed in busy, stressful environments, it can become part of the "background noise" and actually distract from, rather than aid in, attainment of situational awareness. Monitor height, distance, and peripheral vision differences need consideration to optimize operational efficiency and prevent strain and injury.

An important lesson has been learned from 9-1-1 Wireless Phase II deployments wherein detained mapping of call origin is displayed to a map for the benefit of the call taker. There is wide variation in an individual's spatial orientation and innate ability to interpret abstract mapping information. Simply put, when driving, some people can look at a map, immediately orient themselves to their present location, and choose the correct path to a destination; others prefer turn-by-turn directions given as the trip progresses. Some PSAP call takers ignore caller map plots and focus exclusively on what the caller is saying, while others glean important location orientation by just a glance at the map.

Data needs to be displayed so it is readily deciphered without cumbersome interpretation or manual calculations. Borrowing from the U.S. Navy's Combat Information Center design, PSAPs should be ergonomically configured, with communications facilitation in mind. The way the data is displayed on the screen is a key factor in the usefulness of the information and the ability of the operator to use it to maximum effect in high-pressure situations.

Intelligent design analytics have been developed to overcome the tendency of missing something as a semi-attentive observer. They use software algorithms to parse the information within video images to determine something out of place, like a suspicious briefcase left in a public lobby. Motion detection can be incorporated with video monitoring to call attention to views needing attention, but areas with regular vehicle and foot traffic make their use problematic, as well as false activation by animals and other harmless background sources.

Data mining on demand

Rather than trying to display all data being delivered with the NG9-1-1 contact session, screen displays should be configured to offer additional information upon demand as situations dictate. The notion of "data mining" allows personnel to see basic information in areas of their display so their primary attention is focused on critical data. In the PSAP environment, the location of the emergency is a primary concern. The top priority of the PSAP in emergencies is to get the proper first responders sent to the right location, so data relative to location should be allocated to prime screen "real estate," with supporting data relegated to minimized, intuitive "drill-down" commands.

Staff protection from disturbing images

One of the initial reactions PSAP personnel had to the idea of receiving video from the public was, "Oh great, it was bad enough that I had to *hear* someone blow their brains out in the past, now I'll have to *watch* it, too?" The introduction of new technology opens fears on the extreme of possibilities. Although the likelihood of any given 9-1-1 calltaker witnessing a suicide is remote, the concern is real, with this specific issue receiving a lot of early attention in discussion circles. PSAP personnel also are periodically harassed, sometimes in sexually graphic ways, by callers able to conceal their origin.

Disturbing or graphic images could produce a wider concern beyond individual workstations. Many PSAPs conduct tours and open houses for members of the public to see the operation and gain greater insights. There is also a possibility that adjacent coworkers handling other incidents would be unduly distracted by shocking images occurring nearby.

Therefore, CIOs need to guard against display of graphic images, yet allow the system in the background to continue recording for reference and evidentiary purposes. Consistent with how broadcast television handles news stories with graphic images, NG9-1-1 systems should be designed with an option for the receiver to "tile" or distort the image to a very general representation, so contact and some awareness can be maintained, yet with any shock, trauma, or disruption to PSAP personnel minimized.

Training

Although some may assume that viewing information on a video screen is a basic skill learned in everyday life and easily transportable to emergency environments, the demanding environment of crisis communication requires specialized training. Even where properly designed video displays are deployed, the success of new technology systems is ultimately dependent on the skill of the person using it. The field of human factors engineering recognizes the importance of considering human design factors in equipment and systems designed to aid workers in a variety of workplaces. It is necessary to consider a wide range of social and operator behaviors when designing technically complex systems, and devise meaningful training sessions for PSAP staffs.

Final thoughts

New systems will present challenges, especially if hastily deployed without proper consideration of pertinent human factors design.

These matters deserve attention in any new NG9-1-1 deployment:

- Recognition of the vast amounts of information available to responders, and the need to display the information in a useful format will require careful design, considering the tendency of people to become overwhelmed with too much information.

- Best use of new data technology requires intuitive design and properly trained personnel.

- Enlightened human factors design considerations need to be matched with procedural, cultural, and technological enhancements to overcome information overload impediments.

- The quality of communications personnel performance predictably decreases in proportion to the amount and types of stimuli and distractions occurring around them.

- Management of distractions is a critical factor for calm, controlled PSAP communications.

CIOs are compelled to seek ways to manage sensory input and distractions with the expected outcome of better communications and decision making in the PSAP.

RONALD TIMMONS is the director of Public Safety Communications in Plano, Texas. The department answers 911 calls and dispatches first responders for the city of 270,000 people, and administers a regional public safety radio system. Ron has 24 years of first responder experience in firefighting and Emergency Medical Services, including service as chief of a New York State fire department. He also has 13 concurrent years of experience as an adjunct assistant college professor. Ron has a bachelor's degree in Fire Service Administration and a master's degree in Public Administration from the State University of New York. Other graduate degrees include a master's in Homeland Security and Defense from the Naval Postgraduate School in Monterey, California, and a Ph.D. in Public Affairs from the University of Texas at Dallas.

REFERENCES

Bazerman, Max. *Judgment in Managerial Decision Making.* 6th ed. New York: John Wiley and Sons, 2006.

Becklen, R., and D. Cervone. "Selective Looking and the Noticing of Unexpected Events." *Memory & Cognition* 11 (1983): 601-608.

Bigley, Gregory A., and Karlene H. Roberts. "The Incident Command System: High-Reliability Organizing for Complex and Volatile Task Environments." *The Academy of Management Journal* 44 (2001): 1281-1299.

Butler, Emily, Boris Egloff, Frank Wilhelm, eds. "The Social Consequences of Expressive Suppression." *Emotion 3* (2003): 48–67.

Gladwell, Malcolm. *Blink: The Power of Thinking. Without Thinking.* New York: Little, Brown and Company, 2005.

Gladwell, Malcolm. *Outliers: The Story of Success.* New York: Little, Brown and Company, 2008.

Hutchins, Susan. G., "Decision making errors demonstrated by experienced naval offices in a littoral environment," in C. E. Zsambok and G. Klein, Naturalistic Decision Making (Mahwah, NJ: Erlbaum, 1997), 207-215.

Klein, Gary A. *Sources of Power: How People Make Decisions.* Cambridge, MA: Massachusetts Institute of Technology, 1999.

Klein, Gary A. *Intuition at Work.* New York: Currency Doubleday, 2003.

Lindblom, Charles E. "The 'Science' of Muddling Through." *Public Administration Review* 19 (1959): 79-88.

McDermott, Rose, Jonathan Cowden, and Cheryl Koopman. "Framing, Uncertainty, and Hostile Communications in a Crisis Experiment." *Political Psychology* 23 (2002): 133-149.

Milord, James T., and Raymond P. Perry. "A Methodological Study of Overload." *Journal of General Psychology* 97 (1977): 131-137.

National Commission on Terrorist Attacks Upon the United States. Eleventh Public Hearing. New York: W. W. Norton & Company, 2004.

O'Conaill, Brid, and David Frohlich. "Timespace in the Workplace: Dealing With Interruptions." In Proceedings of Computer-Human Interaction '95, 262–263, Denver, Colorado: ACM Press, 1995.

Rensink, Ronald A., J. Kevin O'Regan, and James J. Clark. "To See or Not To See: The Need for Attention to Perceive Changes in Scenes." *Psychological Science* 8 (1997): 368-373.

Roberts, Karlene H., and Denise M. Rousseau. "Research in Nearly Failure-Free, High-Reliability Organizations: Having the Bubble." *IEEE Transactions On Engineering Management* 36 (1989): 132-139.

Simon, Herbert. *Models of Bounded Rationality, Volume 3, Empirically Grounded Economic Reason.* Cambridge, Massachusetts: The MIT Press, 1997.

Simons, Daniel J., and Christopher F. Chabris. "Gorillas In Our Midst: Sustained Inattentional Blindness For Dynamic Events." *Perception* 28 (1999): 1059-1074.

Speier, Cheri, Iris Vessey, and Joseph S. Valacich. "The Influence of Task Interruption on Individual Decision Making: An Information Overload Perspective." *Decision Sciences* 30 (1999): 337-360.

Speier, Cheri, Iris Vessey, and Joseph S. Valacich. "The Effects of Interruptions, Task Complexity, and Information Presentation on Computer-supported Decision-making Performance." *Decision Sciences* 34 (2003): 771–797.

Timmons, Ronald P. "Radio Interoperability: Addressing the Real Reasons We Don't Communicate Well During Emergencies." Master's thesis, Naval Postgraduate School, 2006.

Timmons, Ronald P. "Interoperability: Stop Blaming The Radio." *Homeland Security Affairs* 3 (2007): 1-17.

Tversky, Amos, and Daniel Kahneman. "Judgment Under Uncertainty: Heuristics and Biases." *Science* 185 (1974): 1124-1131.

Tversky, Amos, and Daniel Kahneman. "The Framing of Decisions and the Psychology of Choice." *Science* 211 (1981): 453-458.

Chapter 13: Geospatial Information Systems (GIS) and Their Evolving Role In Emergency Management and Disaster Response

BY ALAN R. SHARK, IN CONSULTATION WITH MEMBERS OF THE NEW YORK GIS COMMUNITY

Introduction

On September 11, 2001, almost immediately after the first plane crashed into the North Tower of the World Trade Center (WTC), the Geospatial Information (GIS) Unit of the Fire Department of New York City, then called the "Phoenix Unit," swung into action. The first map they produced was simple: It delineated the buildings within the World Trade Center complex and identified them by name and by street address. This proved a valuable guide to the thousands of fire fighters and other emergency personnel converging on Lower Manhattan to conduct rescue operations, many of whom were not familiar with the area. Phoenix staff then superimposed a 75' x 75' grid across a map of the disaster area to support intensive and methodical search efforts. These initial maps provided the first instance of a "common operational picture," or COP, for New York City leadership and responders.

They were only the first of hundreds of mapping products that were produced during the three months immediately following the WTC attack to satisfy the thousands of information requests made by the dozens of federal, state, local, and private agencies that comprised the response community. Within one week of 9/11, even with the collapse of the building that housed the city's Emergency Operations Center and the evacuation of the city's GIS headquarters office two blocks north of the WTC, there were at least six functioning GIS production centers, providing information and mapping support for rescue, response, and recovery operations, and to the public by making maps available over the Internet. It can truly be said that data organized, analyzed, and visualized on the basis of location was *the* predominant way that information was operationalized and made available to response managers. In retrospect, it can be reasonably generalized that effective use of GIS is a pre-requisite for the effective management of a large-scale disaster.

In 2008, the National Academy of Sciences published a study of the GIS role in disaster response:

> "The committee's central conclusion is that geospatial data and tools should be an essential part of all aspects of emergency management—from planning for future events, through response and recovery, to the mitigation of future events." The report goes on to

say: "In all aspects of emergency management, geospatial data and tools have the potential to contribute to the saving of lives, the limitation of damage and the reduction in the costs to society of dealing with emergencies."[1]

A book published by GIS software leader ESRI in 2002, based in part on 9/11 GIS operations states:

"...now, as never before, GIS technology has become integral to any comprehensive disaster management plan—as essential for dealing with a catastrophic event as bandages and radios."[2]

While well trained, led, and equipped first responders are always the most important part of a disaster response, timely and accurate information and intelligence is probably their most important pillar of support.

What is geo-enabled information?

Practically all information related to events and activities in the real world are characterized by one or several types of location data. For computerized information, this means that almost every digital record has one or more fields that capture either an address, a geospatial coordinate (latitude/longitude), a parcel identification number, or some other indication of where an object or an event is located. While different databases may contain unique assemblies of hundreds of other attributes, when a jurisdiction standardizes on a common set of addresses and coordinates, it is then possible to use ubiquitous location fields to link information together across databases. Moreover, GIS then makes it possible to visualize these combinations of data in map form and to produce a wide variety of analytic products, such as flood risk and plume analyses.

For emergency managers, the importance of location can most immediately be understood in reference to E-911 emergency response operations. Critical to the dispatch of police, fire, and EMS personnel is knowing not only what kind of incident is occurring, but ***where*** it is taking place, ***where*** the closest response resource is located, and what is the ***best route*** to get there, taking into account one-way streets, street closings, and road work. Dispatch systems are designed to manage location information in the fastest and most accurate way possible so that response times are kept at a minimum—and it is widely understood that the faster the response time, the greater the chance for saving lives, minimizing injury, and reducing property damage. When dispatch information is imprecise or downright wrong, there are documented instances where this has directly led to the death of someone in need of rapid assistance. Getting location right is a very, very serious business.

A related example is relevant for homeland security managers concerned about infrastructure protection. One Call (811) operations exist in cities and counties across the U.S. to coordinate street openings. One Call systems require that as part of the permitting process, a property owner, utility company, or public works agency wishing to excavate be-neath the street or sidewalk surface must first ascertain where all buried infrastructure is located, to ensure that lines are not cut or damaged. Agencies and companies in charge of underground utilities examine their maps and engineering drawings in the vicinity of the proposed excavation and then send workers into the field to mark out locations on the street with spray paint. In more sophisticated operations, the utility lines can be assembled electronically onto a common map for use in the field. One Call enables utilities to avoid outages that leave their customers without service, and reduces the high cost of repairing broken infrastructure. Injuries and deaths also can be avoided. In a large city or urban county, there may be many dozens of One Call dispatches daily.

What applies to E-911 and One Call operations is relevant to just about every kind of government initiative that delivers a service to the public or supports a jurisdiction's physical plant. It is no surprise, then, that disaster-related operations—which can be seen as a combined E-911 and One Call response only on a vast scale—similarly require the highest quality location information delivered in close to real time. Disaster response, if properly planned, can leverage the geospatial data and technologies used for innumerable steady state applications.

Origins of geospatial data and enterprise systems

Typically, an enterprise GIS system is based on aerial photography, which produces an accurate image or picture of an entire jurisdiction with each point (pixel) on the image-map being within a foot or less of absolute accuracy. Using the imagery layer as a base, features are extracted and formed into layers that can include streets, building footprints, parcel boundaries, water bodies, elevation, land use, infrastructure networks, population, and service boundaries. Individual agencies using these "common" layers as a foundation can build additional customized layers corresponding to their own responsibilities and operations. With their geo-data in hand, agencies then can design applications to support their operations. With a large base of common data layers and shareable program components, enterprise GIS systems facilitate the development of applications across many agencies. GIS systems have been shown to pay for themselves within three to five years. Improved tax collections made possible by GIS analysis and operations support can by themselves justify the building and sustaining of an enterprise GIS system costing many millions of dollars.

Approaching enterprise data integration: Agencies within a jurisdiction that has an enterprise GIS are strongly motivated to exchange strategic data layers, and can derive great benefit from sharing and integrating information. Because agency map layers are built on a common enterprise-wide base, these layers can be exchanged between agencies with the assurance of a high degree of accuracy and compatibility. As geospatial layer-sharing increases, it can start to approach true "enterprise data integration:" an information technology goal from the inception of the technology revolution. In a jurisdiction with an enterprise GIS, the emergency manager will be able to import data layers built by any agency, with the assurance that the acquired information meets common accuracy and compatibility standards.

Sharing and aggregating data beyond the jurisdiction: When adjoining jurisdictions working with their state adopt common geospatial data formats and establish data-sharing protocols, geospatial information can be combined across government boundaries—much the same way that standardized LEGO pieces can be assembled—leading to ever greater benefits and opportunities for collaboration. This has particular relevance during a disaster event, which rarely involves just one government entity. Responder teams may need to be called in from the surrounding region or from across the nation. Supplies may need to be transported via air, rail, truck, or ship across thousands of miles. Disaster victims may require transport to distant hospitals and shelters. Infrastructure damage at the center of the event may affect the surrounding regions. In each of these instances, combining the geospatial information across multiple states to relate and make visible all these moving parts and essential facilities is important to the success of the response.

The information challenge of a disaster

Disaster prevention, planning, and response require extraordinary expertise, anticipation, collaboration, and courage. Because disasters often occur by surprise, evolve with great rapidity, affect large areas, and cause widespread damage, information systems to support emergency operations must be capable of responding quickly, accurately, and comprehensively. A jurisdiction that has developed good GIS information and has brought the assets together in its EOC is in an excellent position to quickly understand the physical features and the numbers of people threatened by an emergency event. Such jurisdictions also have the information foundation to support the disaster response. Therefore, GIS operations must be integrated into the Incident Command System (ICS) developed by the Federal Emergency Management Agency (FEMA). If the lead agencies of emergency support functions (ESF) adopt a common GIS foundation, internal ESF operations will be enhanced, and cross functional support and collaboration will be greatly facilitated.

Preparing and sharing pre-incident data is only one dimension of GIS disaster response. Data collection efforts also must extend to the new information generated by the disaster itself. A disaster typically causes violent changes to the natural and the built environment that can continue and evolve for days and weeks. Because infrastructure systems are so interdependent, damage to a critical feature may set off a series of even more dangerous cascading effects. People across a wide area may be injured, cut off from essential services, and in need of immediate assistance. Successful emergency operations require the rapid and continuous monitoring and collection of information that identifies these changes and enables responders to react quickly to new conditions. The requirement to be aware at all times of all aspects of a disaster event is often expressed as the need for a common operational picture (COP).

What really is a common operational picture (COP)?

The sharing of critical information among response agencies, between first responders in the field and incident managers at the EOC; across jurisdictions, states, regions, and even countries; and between all levels of government and the private sector is understood to be of vital importance during a disaster response. Shared information is critical for situational awareness, which is the basis for all decision making. (How can you make a good decision without accurate and current information to base it upon?) Developing a common operational picture (COP) is the task most often associated with the bringing together of geospatially enabled information so that all key responders can see the same information and better coordinate their efforts and collaborate.

COP, with its conceptual origins in military doctrine, is most often tied to the production of maps showing the locations of friendly and enemy troops across a battlefield.[3] As applied to domestic disaster response, the National Incident Management System (NIMS) says: "A common operating picture is established and maintained by gathering, collating, synthesizing, and disseminating of incident information to all appropriate parties involved in an incident." NIMS goes on to state that achieving COP allows on-scene and off-scene personnel to have the same information; and enables the incident commander (IC), Unified Command (UC), and supporting agencies and organizations to make effective, consistent, and timely decisions. It states that COP information must be updated continually to maintain situational awareness.[4]

We should be looking at the COP as one very important component of the overall geospatial information environment required to deal with a disaster.[5] In a disaster, each responding agency likely will have its own missions, each with its own information needs. The Disaster Geospatial Information Environment (DGIE) must be designed in advance so that all the entities in a response get the information they require while contributing information generated by their observations and activities. If all this information was included in one centralized COP, the "picture" would be an overwhelming amount of detail that no one could possibly make sense of. The central COP must be an ever-changing selection of the information judged to be most important for sharing among incident leaders to support strategic decision making, and facilitate collaboration and coordination between response partners.

A diverse team of information and incident specialists having the broadest possible view of the overall disaster information environment needs to select and continually update the key data components of the central COP so that it keeps pace with the ever-evolving character of the disaster. More narrowly focused agency and mission-specific COPs also need to be developed, often with greater technical detail and complexity. The details of the DGIE and the master COP and mission-specific COPs that emerge from it still need to be fully defined and are beyond the scope of this chapter. But, bearing these concepts in mind, we can now go on to look at important elements of a comprehensive geospatial disaster response.

Building a comprehensive geospatial disaster response

Emergency managers have an enormous number of geospatial capabilities at their disposal from local, state, federal, and private partners. The challenge will be to identify, select, acquire, and integrate them to meet your jurisdiction's needs and requirements. One size will not fit all. There are a number of key strategies that bear thinking about.

Networking for geospatial effectiveness: The effectiveness of information technology is diminished when data is kept within agency stovepipes and in formats that make integration difficult. On the other hand, the value of information systems increases when data from diverse sources can be integrated easily and rapidly turned into needed products. Enterprise GIS, when done correctly, creates the potential for all organizations within a jurisdiction or a wider region to share data in any combination necessary to solve a problem or support an operation. Improved sharing methods and tools continue to be made available. For example, many states currently maintain public data warehouses that allow the downloading of large selections of state and local information. Additionally, an increasing number of organizations now deploy web mapping and feature services (WMS, WFS) that enable authorized users to browse a remote site, search for a desired information layer, and either view or download that data into their own environment.

As the technical means for accessing data improve, a question comes to the forefront: How can data owners from different organizations facilitate sharing and collaboration especially when the information in question is considered sensitive? Relationships between agencies and governments, necessary for cooperative and coordinated action, don't materialize out of thin air. Every organization has information they are reluctant to share for security and competitive reasons. Sometimes it is precisely this closely held information that is most critical to share—even if in a limited way—with the response community. Among the most important pre-requisites for effective geospatial support for your EOC is the existence of a network of GIS practitioners in your region who are willing to work together to form a trusting community and work out sharing arrangements. Happily, most states and large jurisdictions in the U.S. have at least one GIS users group or network. Emergency managers should make certain that their GIS staff actively participates in these groups and uses them to build partnerships. Data sharing is critical to all phases of emergency management but is most essential during a response when all those trust relationships, developed in advance, enable rapid data sharing when it counts most. No one should need to engage in protracted, time-wasting negotiations for data when lives are on the line.

Assess your GIS capabilities: Arrange to have one or several extended meetings with your jurisdiction's GIS Director and with other key GIS leaders. Let them lay out for you the data, applications, and technologies that already have been created for the EOC and for other response agencies within your jurisdiction. Those assets must then be related to emergency management needs, including all the emergency support functions. If there is a noticeable absence of a particularly important and strategic set of data layers, or other data preparedness gap, you must find ways to meet these needs. For example: If your jurisdiction is at frequent risk of storm surge, it is essential that you have accurate land elevation data and shoreline bathymetry to do accurate flood modeling and to develop evacuation plans. There are many programs at the federal and state level that you can explore to help you achieve your data-building goals. Your GIS capabilities also need to be related to the emergency management software that you have selected to support your operations. ETeam, WebEOC and DisasterLAN are applications that support disaster operations. These systems can interface with your GIS operations and can incorporate GIS map products. Determine the degree to which your EM software is interactive with your jurisdiction's GIS and then decide whether a stronger integration of the two would add value.

Prepare ahead of time: As with so many aspects of emergency management, the worst time to collect the basic information necessary for a swift and effective response is in the midst of the chaos of a major event. Assuming your jurisdiction has a robust GIS data infrastructure, it will be necessary for your EOC to have ready-to-use critical information, including imagery (overhead and oblique), the transportation network, building footprints, parcels, ground elevation, water features, etc. Many EOCs have developed data catalogues that describe each of the data layers they keep on hand, or that are readily available from other agencies and organizations. Additionally, it is essential that key analytic layers be produced ahead of time, including high-quality flood and storm surge risk information. Based on the threat profile of your jurisdiction, other kinds of predictive GIS analyses can include plume models for your most potentially dangerous chemical storage and manufacturing plants. The FEMA HAZUS application can be used to model in advance the type of damage that might be expected from a major storm or earthquake, and the kinds of resources that would be needed to respond effectively.

Mapping templates of the GIS products most likely to be needed during an emergency also should be prepared in advance to save time. For example, following a hurricane or other high wind event, it is essential that the status of key roads be rapidly evaluated and communicated. Within a map template, vital roads and transportation hubs should be depicted in advance, with various status icons at the ready to identify where debris, damage, or high water makes travel difficult or impossible. This template then can be rapidly populated with field information and quickly modified when road status changes.

Use GIS to maintain vigilance: Watch Command serves as an EOC's early warning system. The sooner a threat can be identified, the sooner you will be able to effectively activate and mobilize your resources. By anticipating problems and pre-positioning personnel, equipment, and information in advance, you can maximize resources available for immediate use during the "golden hours," giving you the greatest opportunity to minimize death, injury, and damage. There are a number of GIS tools that can give you a jump on disaster preparedness and response, and these must be at the fingertips of your Watch Command personnel, who should be trained in their use.

For example, Watch Command must be able to track weather events using the National Oceanographic and Atmospheric Administration's (NOAA) National Weather Service's (NWS) online map viewer that can predict the path and the intensity of major storms, such as hurricanes and blizzards, often days in advance (http://www.weather.gov/). The NWS also can identify areas, which due to excess snow accumulation or rain fall, are vulnerable to flooding. Real-time measures and predictive estimates of river height and flood risk are available for viewing or download and can be combined with your own property and elevation maps. The Army Corps of Engineers maintains a website that monitors ice jams and provides historic data of past ice jams to alert jurisdictions of this flooding risk. (See: http://www.crrel.usace.army.mil/ierd/icejam/icejam.htm) The United States Geological Survey (USGS) provides historic and real-time map data on seismic events and tsunami risk. (See: http://earthquake.usgs.gov/earthquakes/recenteqsww/ and (http://nctr.pmel.noaa.gov/) Similar tools exist for wildfires (http://activefiremaps.fs.fed.us/), drought (http://www.drought.gov/portal/server.pt/community/drought_gov/), and communicable diseases such as the flu (http://www.cdc.gov/flu/weekly/usmap.htm), among others. The GIS-based predictive and tracking tools most suitable for the threat environment of each EOC should be identified and put into active use. Watch Command will need the support of GIS staff to get up to speed on the most appropriate sites for your jurisdiction.

Respond with maximum information support: Should a disaster event actually occur, your GIS support operation will need to flip quickly into high gear. Everything will depend on whether you have properly planned to scale up your GIS operations to meet the extraordinary information needs of a disaster.

- **Networks of EOCs exchanging COPs:** A jurisdiction's Office of Emergency Management generally operates the "Central Command" EOC. However, it is almost certain that a number of operating agencies will need to activate their EOCs, as well. Also likely to be involved are the state EOC, federal agency EOCs, EOCs of jurisdictions across the region, and private sector EOCs. Each EOC will be the hub of a significant disaster response operation. Exchanging essential components of the COP between EOCs will be vital to ensuring the response is properly coordinated. How the different EOCs communicate needs to be designed in advance. Emergency planners must take into account that different EOCs will have different GIS viewers, data formats, and emergency management software packages. All barriers to the rapid sharing and integration of essential information must be reduced or eliminated.

- **Expanding capacity:** To manage the information needs of a large scale disaster, the Central Command EOC must be able to rapidly multiply the number of trained GIS personnel and have sufficient workstation, plotters, and data storage capabilities on hand. In New York City prior to 9/11, the Office of Emergency Management had one part-time GIS analyst. By the end of the first week of the response, NYC OEM alone had more than 50 GIS analysts working on shifts that provided 24 X 7 coverage to meet the information demands of the response community. It was fortunate that the hundreds of members of NYC's GIS community were highly networked through the NYC Geospatial Information and Mapping Organization (GISMO) and were able to be rapidly mobilized. GISMO members came from government, the private sector, not-for-profit organizations, and colleges and universities. Many served for weeks at the EOC as unpaid volunteers. All EOCs must have the ability to rapidly ramp up all aspects of their GIS operations to meet the needs of a disaster event.

- **Collecting and organizing field information:** A major disaster requires the deployment to the field of units from a number of different agencies with specialized missions. EMS, fire, and police teams will conduct rescue operations and work to suppress dangerous conditions. Public works teams will assess damage and start to repair water, sewer, transportation, and related infrastructure. Public health teams will identify health risks, and might be required to distribute medicine and vaccinations. Property department teams will assess damage to buildings and identify dangers that might require evacuation or repair. Utility teams will look for network breaks and outages, and seek to restore services as rapidly as possible. These field workers need their mobile devices to have access to key data, ranging from the location of disabled individuals to the structural composition of buildings and the presence of hazardous materials. Because first responders have direct contact with the evolving effects of the disaster, they also have the potential to be eye-witness informants. If their field devices are properly equipped with global positioning sensors and cameras, they can provide invaluable geocoded intelligence, in real time, across the entire landscape of the disaster—the coordinates of what could be hundreds of inputs automatically arranging themselves accurately on a central map. Also, having the GPS position of all field crews maintained in real time on a common map helps to identify opportunities for mutual aid between teams working in proximity to one another who might otherwise be blind to each others' presence. Additionally, GIS systems already have been used to organize "crowd sourced" data from citizens caught in the disaster area.

- **Integrated viewers:** A new generation of GIS viewer is currently in the process of being developed that will make COPs richer in content and easier to exchange. Such viewers can receive feeds from remote sensing devices, unmanned aerial vehicles (UAV), video cams, first responders in the field with handheld devices, cell phones, and data feeds from web mapping and feature services. These viewers also can display multi-layered floor plans and engineering drawings in 3D. The COPs created by these viewers can be made available to the entire response community, from emergency response teams in the field to the NICC and NOC in Washington, D.C. While still in the development phase, these viewers have the potential to revolutionize the way disasters are managed. One example can be found at the following link: http://www.vcoresolutions.com/fourdscape.php. This system is on display at the Long Island Forum for Technology in Bethpage, Long Island. Product development and commercialization are being supported by New York State and by the DHS Science and Technology Directorate.

State geospatial resources

Many state emergency operations centers have substantial GIS capabilities that during a disaster should be integrated with local EOC efforts. Of course, these collaborative efforts need to have been worked out in advance. Following the 9/11 terrorist attack on the World Trade Center, the New York State GIS Division, based in Albany, which had personnel with remote-sensing expertise, coordinated most of the aerial photography operations taking place in the skies over NYC. State remote-sensing support provided essential information and removed a major burden from strained NYC GIS personnel. Luckily, prior to 9/11, NYS and NYC GIS managers had established a strong collaborative relationship, which paid off during the crisis.

Post-9/11, a number of states have developed disaster-related mapping systems that during steady state and emergency conditions can be utilized by local jurisdictions. Several years ago, NYS GIS deployed the Critical Infrastructure Response Information System (CIRIS) that continues to provide state and local responders with online access to hundreds of layers of critical infrastructure information.

Because the effects of a major disaster cannot be confined to one jurisdiction, states are essential partners capable of bringing together information and analysis from multiple jurisdictions in the surrounding region where ripple effects from the disaster will be felt. Local EOC and GIS managers need to sit down with their state counterparts and identify in advance geospatial capabilities and networks that would be shared in case of a major event.

Federal geospatial resources

Federal government agencies, including the Departments of Homeland Security, Interior, and Defense, among others, recognize the importance of geospatial systems for disaster and emergency response, and have been developing valuable geo-oriented data, analysis, technologies, and techniques. However, this wealth of capabilities doesn't automatically translate into improved operations at the state and local level. It is the job of state and local EOCs and public safety agencies to examine these offerings, work with their federal partners, and acquire, customize, and integrate those capabilities that meet their needs. It helps that, over the past few years, federal agencies have been increasing their regional outreach activities. The following are a few of the offerings that state and local governments can draw upon. In the space allowed, only brief descriptions can be offered. Follow the references and URLs provided to find additional information.

GeoCONOPS: The GeoCONOPS is a multiyear effort focused on the geospatial communities supporting DHS and FEMA activities under the NRF. The Federal Interagency GeoCONOPS is intended to identify and align the geospatial resources that are required to support the NRF, ESF, and supporting mission partners. The GeoCONOPS document has many relevant sections that state and local governments will find of value. Go to http://www.napsgfoundation.org/blog/napsg-blog/113-dhs-releases-geoconops-30-a-napsg-call-for-input to download the Final Draft of Version 3.0 of the GeoCONOPS, dated June 2011. Many of the descriptions of federal capabilities in the following section are drawn from the GeoCONOPS publication.[6]

Homeland Security Information Network (HSIN): The Homeland Security Information Network (HSIN) is a national secure and trusted web-based portal for information sharing and collaboration between federal, state, local, tribal, territorial, private sector, and international partners engaged in the homeland security mission. HSIN is made up of a growing network of communities, called Communities of Interest (COI). COIs are organized by state organizations, federal organizations, or mission areas such as emergency management, law enforcement, critical sectors, and intelligence. Users can securely share within their communities or reach out to other communities as needed. HSIN provides secure, real-time collaboration tools, including a virtual meeting space, instant messaging, and document sharing. HSIN allows partners to work together instantly, regardless of their location, to communicate, collaborate, and coordinate. http://www.dhs.gov/files/programs/gc_1156888108137.shtm

GeoSpatial Information Infrastructure (GII): The DHS Geospatial Information Infrastructure (GII) is the governing body of geospatial data and application services built to meet common requirements across the DHS mission space. DHS OneView is a lightweight Internet application providing geographic visualization and analysis to individual users. OneView is implemented within the GII by the Geospatial Management Office (GMO). OneView provides access to HSIP Gold data layers and to other federal data resources. OneView users have the ability to add external data sources to their view in common web service formats (KML, KMZ, WMS, and GeoRSS). Anyone with a valid HSIN account may access OneView at https://gii.dhs.gov/oneview.[7]

HIFLD Working Group: The Homeland Infrastructure Foundation-Level Data (HIFLD) Working Group was established in February 2002 to address desired improvements in collection, processing, sharing, and protection of homeland infrastructure geospatial information across multiple levels of government and to develop a common foundation of homeland infrastructure data to be used for visualization and analysis on all classification domains. The HIFLD website (https://www.hifldwg.org/) contains extensive information, useful links, and application forms related to HSIP Data and HIFLD activities. If you work for government in a HLS/public safety capacity, you can apply for HIFLD access.

HSIP Gold and HSIP Freedom: HSIP Gold is a unified homeland infrastructure geospatial data inventory for common use by the federal Homeland Security and Homeland Defense (HLS/HD) Community. It is a compilation of over 450 geospatial datasets, characterizing domestic infrastructure and base map features, which have been assembled from a variety of federal agencies and commercial sources. HSIP Freedom is a subset of HSIP Gold data that can be made available to state and local governments. https://www.hifldwg.org/

Virtual USA (vUSA): Virtual USA, an initiative of the DHS Science and Technology (S&T) Directorate, is creating a new way for emergency managers to operate and collaborate. Virtual USA was built by a partnership of practitioners who defined:

- how shared information should be documented and described,
- the technical tools necessary to share information,
- agreements needed to set standards for sharing,

- the partners that should join the sharing consortium, and
- how access to the system would be controlled.

Virtual USA gives emergency managers transparency into the information that is shared, as well as control over who has access to it, where, and for how long. All shared information lists a personal point of contact, so users know who to call with questions. https://vusa.us.solution.html

Interagency Remote Sensing Coordinating Council (IRSCC): As stated in the Federal Interagency Geospatial Concept of Operations (GeoCONOPS): The Interagency Remote Sensing Coordination Cell (IRSCC) is an interagency body of remote-sensing mission owners with capabilities that enable the primary federal responder to plan, coordinate, acquire, analyze, publish, and disseminate situational knowledge. When activated, the IRSCC provides visibility of the remote-sensing missions that are the statutory responsibility of the member organizations. This provides the community (federal, state, local, and tribal governments) information about ongoing remote-sensing missions before, during, and after a Stafford Act declaration.[8]

Next steps, key considerations

As you move forward in your thinking about how you wish to take advantage of geospatial data and technologies to support your emergency management and disaster response operations, you may wish to keep in mind the following objectives:

1. **Work toward defining your Disaster Geospatial Information Environment (DGIE), central COP and mission COPs:** Choreographing the movement of data to get the right information, into the right hands at the right time is key to a successful disaster response, but it is a complex and demanding effort. It will be necessary to understand the information needs of all potential disaster responders and align them with information sources ranging from local databases to remotely sensed data captured by satellites, fixed-wing aircraft, and UAVs. Each agency likely to participate in a disaster response should be asked to reveal its disaster information management plan: how it will collect data from the field, how it will produce analytic and operational products, and how it will create its own, customized COP. Every jurisdiction's EOC needs to develop a methodology by which a central COP can be formed through the selection of the most important information inputs from across the response community. This is likely a long-term design and development effort that will require constant testing and fine tuning.

2. **Put data sharing agreements in place to prevent negotiations during a disaster:** Sharing data between agencies, between levels of government, and between private firms is not easy, but in a large-scale disaster such sharing is essential to provide responders and the public with critical information, which likely will affect their safety and well-being. Without knowing the status of electric outages, the work of many gov-

ernment agencies may be impeded. Without knowing the condition of roadways, crews seeking to restore power may find they cannot reach downed wires or damaged transformers. The Virtual USA program (vUSA) of the DHS Science and Technology Directorate can help to promote data sharing between organizations that reduces or eliminates concerns about the exposure of sensitive data. Another federal capability that can help promote data sharing is the Protected Critical Infrastructure Information (PCII) program. PCII (http://www.dhs.gov/files/programs/editorial_0404) is an information-protection program that enhances voluntary information sharing between infrastructure owners and operators, and the government. PCII protections mean that homeland security partners can be confident that sharing their information with the government will not expose sensitive or proprietary data. The vUSA and PCII programs should be considered when designing your data-sharing plans.

3. **Develop GeoCONOPS and GeoSOPS that reflect your jurisdiction's unique threats and vulnerabilities:** Following the 9/11 terrorist attack on the World Trade Center in NYC, the GIS community was asked by NYC OEM to develop maps and analysis to support the response. The Emergency Mapping and Data Center (EMDC) had to start from scratch. They didn't know what elements needed to be put into the maps demanded by incident commanders and the mayor. They did not know what information products responders required to best support their efforts. They did not know how to effectively collect critical intelligence from the field. They didn't know how to properly distribute the maps produced. They had to learn all these things, and many more, on the fly with considerable waste of time and effort. Emergency managers should consider building upon the GeoCONOPS work already done by federal DHS, and develop detailed standard operations procedures covering information collection and analysis, COP development, and information distribution activities. Never neglect an opportunity to bring GIS into your exercises to the maximum extent possible. During a disaster, GeoCONOPS and GeoSOPS will ensure that GIS personnel know exactly what they need to do, so that no time will be wasted producing those first essential maps.

4. **Take responsibility for integrating all available tools:** A great wealth of GIS data, GIS technology and federal geo-oriented products and capabilities await your use. Matching all these options with your particular challenges is a difficult undertaking. It will be up to each region, each state, and each locality to integrate the mix of capabilities most suited to their needs. Fortunately, this does not need to be done in a vacuum. State and local EOC managers can consult with a variety of federal GIS personnel in their region. EOC managers also can work with national GIS organizations whose mission is to foster collaboration across states, counties, cities, and municipalities. These organizations include the National States Geospatial Information Council (NSGIC: www.nsgic.org), Public Technology Institute (PTI: www.pti.org), the National Alliance for Public Safety GIS Foundation (NAPSG: www.napsgfoundation.org), and the Geospatial Information Technology Association (GITA: www.gita.org). These organizations are working to identify best practices and to promote collaboration and sharing. They also can help put you in touch with the federal GIS representatives in your region.

Conclusion

We are at the leading edge of a new information-based revolution in the way disasters are managed and in the way critical data is generated and exchanged within the response community and with the public. Thanks to the efforts of government and the private sector, the lessons learned from the geospatial responses to past disasters have led to new capabilities better able to deal with these disruptive events. The visualization, integrative, and analytic powers of GIS make it ideally suited to support the needs of emergency responders from incident commanders to rescue workers in the field. If we continue to work collaboratively, our progress will be rapid, and the payback can be enormous.

DR. ALAN R. SHARK is credited writing this chapter with the active consultation from the local New York GIS Community. Dr. Shark serves as the Executive Director and CEO of Public Technology Institute (PTI). Celebrating its 40th year, PTI is a national, non-profit organization that focuses on technology issues that impact local government and thought-leaders in the public sector. Dr. Shark's career has spanned over 28 years as a highly recognized leader in both the nonprofit management and technology fields, with an emphasis on technology applications for business and government. He is an associate professor of practice at Rutgers University where he teaches a masters-level course on technology and public administration. Dr. Shark was elected a Fellow of the National Academy of Public Administration. He received his doctorate in public administration from the University of Southern California.

ENDNOTES

1. Successful Response Starts with a Map: Improving Geospatial Support for Disaster Management, National Academy of Sciences, 2007, ISBN:978-0-309-10340-4

2. Confronting Catastrophe, A GIS Handbook; page X, R.W. Greene, ESRI Press

3. Wikipedia; available at http://en.wikipedia.org/wiki/Common_operational_picture; Internet; accessed May 7, 2012

4. Department of Homeland Security, "National Incident Response Plan," December 2008, Pages 23-24; www.fema.gov/pdf/emergency/nims/NIMS_core.pdf

5. Lieutenant Colonel Jeffrey Copeland, "Emergency Response: Unity of Effort Through A Common Operational Picture," U.S. Army War College, Carlisle Barracks, PA 17013-5050; www.dtic.mil/cgi-bin/GetTRDoc?AD=ADA479583

6. Department of Homeland Security, Federal Interagency Geospatial Concept of Operations (GeoCONOPS), Final Draft, Version 3.0, June 2011

7. Ibid: pages 153-154

8. Ibid: pages 151-152

Chapter 14: Spatial Intelligence in Public Safety

BY PAUL CHRISTIN

Introduction

The public safety mission is very demanding, challenging, unpredictable, yet often rewarding. The majority of people choosing a career in public safety—whether it is law enforcement, firefighting, or emergency medical services—don't do so for the money; knowing that they helped someone in need motivates them most. When someone begins a career in the public safety service, they typically do not think they will be using advanced technology other than the tools of their trade. Very few probably know how valuable one particular technology would be to them in the process of handling their daily tasks of protecting their community. The technology? Geographic Information Systems (GIS).

Using maps in public safety is not a new concept. Historically, paper-based maps were used to track down criminals on the run, to identify terrain features, to identify streets and addresses, and so on. However, the modern age has brought computer technology to public safety, and mapping is no exception. Geographic Information System (GIS) technology is increasingly becoming ubiquitous throughout public safety. Whether for a first responder, commanding officer, or chief executive, GIS technology is providing spatial intelligence for effective decision making.

Most public safety professional use GIS every day and do not even realize they are dependent on the technology. From the moment an emergency call occurs, GIS is hard at work, locating the caller, looking for the closest available unit, and then providing routing directly to the scene with turn-by-turn instructions presented on a map.

Throughout this chapter, we will discuss how GIS supports the public safety mission, how to implement GIS technology, and a vision for future implementations.

GIS in public safety today

Many communities across the globe provide their residents with a three-digit number to dial when an emergency occurs and help is required. Regardless if it is from a landline or mobile phone, when an emergency call is made, the caller's location is identified, and the information is passed to an emergency call taker and then to an emergency dispatcher. GIS is important to serving the public safety mission beginning with an emergency dispatch center.

The primary system managing emergency call handling is a computer-aided dispatch system (CAD). A majority of CAD solutions available on the market provide GIS capabilities. They vary in the depth of their integration. Some simply use GIS to plot incidents on a map. Others have developed on a GIS platform to provide comprehensive geospatial analytics and functions throughout the dispatching workflow. Using GIS, emergency dispatchers easily identify the caller's location or identify the location of automatic alarms, see the location of resources using automatic vehicle location (AVL) systems, and query a records management system for premise history. Dispatchers communicate relevant incident information to resources in the field through radio communication, or information automatically transfers to mobile data terminals in vehicles, providing better and safer responses to emergencies.

Beyond the emergency dispatch center, GIS technology provides law enforcement, fire/rescue, and emergency management operations with geospatial intelligence. Law enforcement agencies rely on GIS technology to analyze criminal activity, conduct investigations, create prevention programs, and evaluate department performance (known as the ComStat process). GIS provides the ability for crime analysts to identify trends, patterns, and behaviors in criminal activity, such as Hot Spot Analysis. Armed with this information, officials can increase law enforcement presence as deterrence in areas affected by a high concentration of activity. Additionally, analysts use GIS in the investigative process to solve crimes that have occurred. For example, an analyst working on a case involving an assault can query offender registry systems to identify individuals with prior history of similar activity and determine their proximity to recent events. Detectives can use this information to perform further investigations to eliminate suspects.

The present threat of terrorism has led to the use of GIS within intelligence centers often operated by law enforcement agencies. Many of these centers intake suspicious activity reports (SARs), assess their significance, and evaluate whether the activity has a nexus for terrorism. GIS is an important tool in the analytical process because of the ability to identify relationships in a geographic context. Using GIS, analysts can identify the proximity of a SAR to critical infrastructure, special events, or other SARs to help determine the legitimacy of the report.

The fire service increasingly is using GIS to perform community fire risk assessments, structural pre-plans, station locations, and fire ground operations and accountability. Many published studies address the subject of fire risk and hazard modeling. However, few have defined the role of GIS technology in assessing community vulnerabilities, or modeled the relationship and interdependencies between hazards, risks, and values.

GIS provides the ability to prepare and coordinate capabilities to prevent, protect against, respond to, and recover from all hazards in a way that balances risk with resources, which is the mission of public safety and emergency management operations.

GIS implementation patterns

As more public safety agencies recognize the value of geospatial intelligence in performing their mission objectives, they are demanding increased GIS capabilities within next generation solutions. As such, it is very important to gain a better understanding of implementing GIS to maximize the desired capability.

There are essentially five implementation patterns to GIS: data management, planning and analysis, field mobility, situational awareness, and citizen engagement.

1. Data management

Richard Saul Wurman is a technical communicator who developed a method for organizing data called the LATCH method: Location, Alphabetical, Time, Category, and Hierarchy. All data can be organized into those five dimensions with the lowest common denominator in data being geography.

GIS provides a platform to fuse disparate data sources based on their spatial associations. Accessing static and dynamic data relevant to location allows users to get various types of information from the map display to meet their public safety mission objectives. Geographical data management provides the ability to apply spatial analytical models and visualize results on a map, enhancing situational awareness in the cloud, over the enterprise, or in the mobile environment.

2. Planning and analysis

Being prepared to respond to any incident largely involves training and a significant amount of pre-planning. GIS provides the ability for public safety agencies to understand the community and environment they serve and protect.

Fire/rescue departments use GIS to assess community risks, incident pre-planning, and during fire ground operations. Spatial models like 3D building models provide fire departments with rapid assessment of buildings reachable with ladder trucks. Other spatial analytics such as "hot spot analysis" provide the ability to improve coverage areas by understanding where there has been a concentration of previous incidents.

Similarly, spatial analytics provide law enforcement agencies insight into where they need to focus crime prevention strategies and tactics. Understanding where there is a concentration of criminal activity allows law enforcement agencies to increase the patrol frequency in those neighborhoods to deter crimes and to improve response times.

Finally, emergency management operations use GIS in all stages of emergency management functions such as prevention, preparedness, mitigation, and response. Spatial analytics provide emergency managers with the ability to assess risks and vulnerabilities associated with natural, technological, and man-made disasters. In addition, emergency managers use spatial analytics to understand the types of mitigation procedures required to protect against catastrophes.

3. Field mobility

A major portion of the public safety mission occurs in the field, on the street, and at incident locations. Having the ability to collect, receive, and share information in a mobile environment is essential. GIS provides the ability in the field to collect and manage data during large incidents, and in everyday operations to easily share with a dispatch or command center for better decision making.

4. Situational awareness

Situational awareness is often synonymous with a common operating picture (COP) and a map. The need to understand who, what, when, and where is paramount to effective incident management. However, there is usually a tendency in public safety to add too much data to the COP, creating a clutter of information, which is difficult to navigate and understand. GIS map viewers configured to be specific to user roles create a clearer display of operational information. Users are better informed and are able to make quicker assessments of the managed situations.

5. Citizen engagement

Many public safety agencies are establishing public portals that use GIS as a means to engage residents to report suspicious activity and crime tips. Public safety agencies also use the portal to notify residents of emergencies and for evacuation notifications, as well as to provide public education on public safety concerns. GIS makes those types of communication more meaningful because residents can put information into the context of where they live, work, and play.

GIS—a common system platform

Many business systems are required to support public safety mission objectives. There are staffing systems to manage rosters and payroll. Dispatching systems support assigning resources to emergency calls for service. Records management system support incident documentation. Case management systems support criminal detectives and prosecutors in building court cases. Lastly, there are specialized systems to manage equipment inventories and other miscellaneous information.

However, many of the systems are disconnected and function in silos, rendering it difficult to access information when it is need the most—at a time of crisis.

GIS is a common platform that helps incident command and first responders:

- Understand their jurisdictions' critical exposures and vulnerabilities

- Identify and prioritize mitigation requirements to prevent/reduce emergency impacts and losses

- Implement and maintain mission-specific situational awareness before and during emergency incidents

- Share data, information, and situational awareness with other organizations, the public, and public safety personnel in the field

- Collect geographically referenced information and data from all types of mobile devices in near real time

GIS functions as a common platform by providing the ability to standardize on a data model such as described earlier using the LATCH method. In addition to a data model and data links to authoritative sources, the platform should include commonly used spatial analytics accessible over any operating environment (desktop, mobile, or within the cloud), and the ability to consume and publish social media.

These geographic platform components allow organizations to conduct comprehensive risk and hazard analysis identifying vulnerabilities, mitigation requirements, response and contingency planning for better preparedness, and comprehensive situational awareness to support better decision making.

The future of public safety GIS

The role GIS technology will play in the public safety mission will become more important than it is today. Computer technology will evolve into systems with artificial intelligence becoming more contextualized based on the user profile and role. Systems will mimic the human experience and provide their users with data and information based on three primary types of human awareness: self, temporal, and spatial. Computers and devices will understand their user's digital DNA and apply analytics and logic derived from inputs obtained from their user's surroundings, time of day, desires, and interests. These innovative concepts will certainly support the public safety mission, making the job safer and more effective.

Spatially intelligent incident systems will provide location information to first responders without user-initiated requests. Upon receiving a call for service, the spatially intelligent system will peer into all databases, connected sensors, video cameras, and available resources and build a dynamic response commensurate with the emergency.

Advanced solutions using Location Based Services (LBS) and Augmented Reality (AR) will be prevalent throughout emergency response. LBS and AR applications will become standard features on the next generation fire apparatus and law enforcement vehicle making them "Smart Vehicles." The smart law enforcement patrol car will project suspect intercept routes on a map through a heads-up display, and it will communicate with traffic control signals to clear traffic from intersections during pursuits. It also will communicate with civilian versions of "smart vehicles" to avoid emergency vehicle accidental collisions.

Augmented reality applications loaded on spatially intelligent devices will provide firefighters entering a structure with floor plans of buildings and help them identify obstructions and hazards just by pointing their device directionally at the exterior of the building. Once inside, LBS applications embedded in "smart buildings" will provide the location of victims trapped inside burning, damaged, or collapsed structures. In addition, the smart building will communicate with the devices carried by personnel to provide information such as the closest exit, alarm panel and pull station locations, first aid stations, hazardous material locations, and sprinkler/HVAC rooms.

Augmented reality also will provide law enforcement officers with the ability to visualize the location of registered criminals or see fleeing suspects captured on video surveillance systems proximal to the scene of a crime. Additionally, officers will be able to pan their devices around to visualize crime data and identify places where crimes are likely to occur based upon predictive analytics.

Historically, the public safety market has been slow to adopt technology—particularly software. However, GIS is proving its value in providing situational awareness within the public safety mission. Many more applications will emerge with similar spatially contextualized capabilities, providing a much safer environment for public safety professionals to operate within. Technology will be a seamless blend between hardware, software, and networks to provide the context for any situation and role. Users will interact with technology in a unique way, as augmented reality is suggesting today. Vehicles, buildings, equipment, and devices become sources for information to interact with, extending the human reach into the digital world, and understanding location and its relevant meaning to mission objectives becomes the primary factor in protecting and securing communities.

PAUL CHRISTIN joined Esri as the Homeland Security Industry Manager in 2006. His primary responsibilities include understanding the critical issues and challenges facing domestic and international homeland/national security missions to apply GIS technology in solving those challenges. He served nearly 15 years in the fire service as a fire apparatus engine/firefighter/paramedic. He is certified in ICS and HazMat operations. Christin has written and published several whitepapers discussing applications for GIS technology in public safety.

Chapter 15: Cyber Security and Information Assurance for Long Term Evolution (LTE)
Public Safety Networks—Today and Tomorrow

BY GREG HARRIS

Abstract

Every aspect of our day-to-day lives is increasingly reliant on a network of some kind. From our personal smartphones the cloud applications at the office, our professional productivity is connected to the networks that connect us. These connections we so depend on, although important, are not critical to our public safety if security is breached and connectivity lost.

If the security of a non-mission-critical communications network is compromised, personal information can be compromised, leading to anger, frustration, and distrust of those networks. For mission-critical communication public safety networks, inadequate security can have much more dangerous and far reaching consequences.

Public safety communications in today's world faces unprecedented challenges. More than ever, homeland security and situation readiness depend heavily on effective communication among federal, state, county, and local agencies. The importance of cyber security (CS) for public safety mission-critical communication systems has only recently been accepted and, therefore, has not yet been formally addressed by agencies recently affected.

Agencies have invested considerable effort in managing the risks of terrorism and other deliberate attacks against their facilities since the devastating events of September 11, 2001. However, these activities have focused mainly on physical security and dealt less with cyber attacks against sensitive information systems. While few deliberately focused attacks on public safety systems have been "publicly" reported, indiscriminate attacks of worms, trojans, viruses, malware, etc. have occurred, and they have adversely impacted information systems.

Cyber attacks against all commercial networks are on the rise in recent years. According to the Identity Theft Resource Center, in 2010 more than 662 security breaches exposed over 16.1 million records.2 And those are just the reported breaches. Significantly, targets for cyber attack will not be limited to commercial networks. The sophistication and frequency of cyber attacks will likely grow over the next few years, and those attacks will increasingly seek to penetrate mission-critical communication networks. If not checked, they could have devastating effects, compromise the missions of first responders and even erode public confidence.

How real is the threat?

- Stuxnet was detected with roots tracing back to June 2009. It has been called the most refined piece of malware ever discovered. Mischief or financial reward wasn't its purpose; it was aimed right at the heart of a critical infrastructure. Symantec estimates that the group developing Stuxnet would have been well-funded, consisting of five to 10 people, and would have taken six months to prepare.[1] It is the first discovered worm that spies on and reprograms industrial systems. It was specifically written to attack Industrial Control Systems (ICS) used to control and monitor industrial processes. Stuxnet could also reprogram the programmable logic controllers (PLCs) and hide the changes.

- In January 2003, hackers released the Slammer worm. It penetrated a computer network at Ohio's Davis-Besse nuclear power plant and disabled a safety monitoring system for nearly five hours, despite a belief by plant personnel that the network was protected by a firewall. This event occurred due to an unprotected wireless interconnection between plant and corporate networks. The Slammer worm had significant impacts on other companies. It took down one utility's critical SCADA network after moving from a corporate network to the control center LAN. Another utility lost its Frame Relay Network used for communications, and some petrochemical plants lost HMIs and data historians. A 911 call center was taken offline, airline flights were delayed and canceled, and bank ATMs were disabled. These were the effects of the release of one *unintelligent* piece of malicious software. No specific facility was targeted.

- tIn September 2001, a teenager hacked into a computer server at the Port of Houston, Texas, to target a female chat room user following an argument. The attack bombarded computer systems used for scheduling at the world's eighth largest port. The port's Web service, which contained crucial data for shipping pilots, mooring companies, and support firms responsible for helping ships navigate in and out of the harbor, was left inaccessible.

Historically, public safety communication systems have been kept separate from corporate networks and the Internet, but increasingly, IT departments are interconnecting them, driven by the need to communicate with separate agency departments, leverage existing corporate infrastructure, and the opportunity to share information through an intranet or the Internet. Public safety systems are exposed to penetration when they are connected to other networks or when they are provisioned for remote access, and existing public safety systems were not designed with the realization of public access in mind. Therefore, they sometimes have poor security, leaving them vulnerable to attack. An historical legacy in the public safety community of treating systems as closed or proprietary has resulted in recently implemented systems that do not contain basic security practices.

Even if security tools (e.g., firewalls, intrusion detection/prevention, antivirus, etc.) are put in place, they are not properly managed as part of an agency's security and risk management plan. Furthermore, much of the technical information needed to penetrate these systems is readily available even as software, called script-kitties, for casual computer users to perform the attack. These issues also affect other types of information systems, including communications, access control, inventory control, power, transportation, and financial systems, all of which are increasingly interlinked.

Who are they, and how will they attack?

Computer systems can be attacked from within an agency or externally, including attacks from outside the country. Data on cyber attacks indicate that about 80% of actual attacks are made by insiders, with 43% of attacks being deliberate malicious attacks, while surprisingly unknowing non-malicious actions cause 60% of the losses.[2]

Attackers include:

- Thrill-seeking, hobbyist, or alienated hackers who gain a sense of power, control, self-importance, and pleasure through successful penetration of computer systems to steal or destroy information or disrupt an organization's activities.

- Disgruntled employees, contractors, or other insiders who damage systems or steal information for revenge or profit. (Even employees who are content in their jobs can inadvertently introduce malicious software into the system from infected mobile devices.)

- Terrorists for whom hacking offers the potential for low cost, low risk, but high gain attacks.

- Professional thieves who steal information for sale.

- "Hacktivists" who use cyber crime as a method of political, economic, and social campaigns for their cause that border on cyber terrorism.

- Criminal organizations that have realized the cost benefit of cyber crime over the risk of physically performing criminal activity.

- Adversary nations or groups who use the Internet as a military weapon for cyber warfare, a discipline the U.S. has already engaged in itself.

Hackers originally were individuals with highly specialized and esoteric knowledge of computer systems. Consequently, they were few in number. However, some of these early hackers made their knowledge available to others through the development and distribution of software packages that provide hacking tools. Some of these packages rival commercial software in their design and are essentially point-and-click applications. A number of them provide suites of hacking tools. Their availability has significantly increased the number of people capable of performing sophisticated hacking. The hacking community spends considerable time probing computer networks for vulnerabilities and will actually publicize them through Internet chat rooms and forums.[3]

The future of Long Term Evolution (LTE)

As public safety agencies and their communications capabilities evolve from Public Safety Land Mobile Radio (LMR) systems to the additional data capabilities provided via LTE, security becomes even more important. Even with private LTE networks, public safety agencies may be relying on external providers for data segregation, data privacy, privileged user access, availability, and recovery. Location independence coupled with the possibility of a service provider as the provider of subcontracted services create risks that go beyond the reach of the typical approach to security.

LTE at 700 MHz will provide high-capacity, high-data-rate coverage over wide areas, minimizing the number of base stations required and reducing costs for hardening and security. It is also uniquely suited to deliver high-data-rate services efficiently and with robust performance. Consequently, LTE can allow public safety field personnel to use bandwidth-intensive data applications such as real-time video, high-resolution still photographs, images, and floor plans, giving them fast access to visual information that can be crucial for responding most effectively to a crisis.

This is especially relevant for an evolved packet core (EPC), which is based on IP transport on all its wired links. While EPC makes the network very efficient, it also requires a comprehensive and verified security architecture to ensure protection against the diversity of threats that jeopardize mobile networks, in particular IP-related cyber attacks.

Solution: Information assurance (IA)

Interoperability requirements and industry standards have driven many industries toward improved and complex security postures as well as enhanced interoperability. The same holds true for networks built with LTE technology. **The solution to address this complex security concern is information assurance (IA). In its broadest sense, public safety IA for LTE addresses the full range of security issues and risks that could affect the day-to-day operations of public safety agencies today and tomorrow.** The requirements of LTE for public safety can make the challenge to provide the highest level of security even more complex. The most common LTE network deployments for public safety will most likely integrate those first responder LTE networks with other private networks, and potentially with commercial networks and LTE carriers. The importance of IA in this type of communications environment is rapidly capturing the attention of public safety agencies across the country and at the federal government level. In fact, IA for LTE networks is being formally addressed by the proposed governing agencies responsible for LTE security.

Public safety agencies and the networks they manage are governed by federal laws that relate to security. The two major laws are the Health Insurance Portability and Accountability Act (HIPAA) and Criminal Justice Information Services (CJIS). HIPPA mandates the privacy and security of medical records, which means that first responder emergency medical services personnel must be able to use applications and devices that can recognize and conform to HIPPA requirements. CJIS has mandates and guidelines for limiting access to criminal justice information and is widely used by law enforcement practitioners.[4]

Security objectives

Secure communications links are vital to the majority of public safety practitioners, together with infrastructure reliability and resiliency. Public safety practitioners may not widely know that security mechanisms are a major component of commercial technologies and their standardization. Specifically, security in LTE can be found in both the user and control planes, either for protecting confidentiality of the information or its integrity.

This discussion recognizes that virtually all networks deployed in the 700 MHz public safety broadband spectrum will adopt LTE technology, specifically to at least 3GPP Standard E-UTRA Release 8 and associated EPC. A significant portion of the security architecture is pre-determined in accordance with the 3GPP standards. Harris Public Safety and Public Communications (PSPC) has adopted the LTE security framework as it relates to the five LTE security groups:

1. Network Access Security

2. Network Domain Security

3. User Domain Security

4. Application Domain Security

5. User Configuration and Visibility of Security

These form the baseline security approach. In addition to the five LTE security groups above, the Emergency Response Interoperability Center (ERIC) Public Safety Advisory Committee (PSAC) report[5] also recommends the implementation of three supplemental security and authentication categories:

1. Roaming to commercial networks

2. Support for varied application and security requirements associated with a diverse public safety market and the applications and software specific to individual cities, counties, regions, and states

3. Access to the Internet

Defense in depth

An industry best practice is "defense in depth." Defense in depth minimizes the probability that the efforts of malicious hackers will succeed. A well-designed strategy of this kind also can help system administrators and security personnel identify people who attempt to compromise a computer, server, or proprietary mission-critical communication network. If a hacker gains access to a system, defense in depth minimizes the adverse impact and gives administrators and engineers time to deploy new or updated countermeasures to prevent recurrence.

Components of defense in depth include antivirus software, firewalls, anti-spyware programs, hierarchical passwords, intrusion detection, and biometric verification. In addition to electronic countermeasures, physical protection of business sites, along with comprehensive and ongoing personnel training, enhances the security of vital data against compromise, theft, or destruction.

The idea behind the defense-in-depth approach is to defend a system against any particular attack by applying just the right amount of security by providing the following capabilities:

- Access control
- Host security
- Physical security
- Centralized logging and auditing
- Intrusion prevention and detection systems
- Encryption key management
- Enterprise backup (disaster recovery)
- Enclave firewalls
- Automated vulnerability management

Taken in its totality, defense in depth is a critical capability. If one security barrier is broken, the next security layer will prevent a successful attack. For example, if an attacker tries an incorrect password, the security layer may provide an alarm/notification to inform personnel that an excess number of passwords have been attempted. In some instances, defense in depth may detect the attack and allow various levels of response to the attack to move into place. This even could result in locking down all access to the affected facilities.

How much security is required?

As those who manage radio systems and networks are aware, the Radio Access Network (RAN) is the most vulnerable to external attacks. With LTE technology and the potential use of more commercially available user equipment, there will be a tendency toward more permitted interoperability in line with the LTE protocols established by the 3GPP LTE security standards. This minimum requirement actually calls for a continuous risk assessment process as part of the ongoing operation of the LTE network in order to respond to newly generated threats.

There is a set of best practices guiding IA for LTE networks for public safety. Specific due diligence is required to balance the cost of implementing security measures against the likelihood and impact of a cyber attack. The cost/impact balance also must recognize the harsh reality that no single security measure is 100% effective in preventing a security breach and that security breaches will inevitably occur. Therefore, layered security measures must be applied, and methods must be developed so that if one security barrier fails, another exists to deter, detect, and cope with the threat, or at least create an audit trail for forensic analysis, possible legal actions, and future training. Therefore, a formal risk assessment is crucial when determining the appropriate levels of security for any public safety communications network. In other words, the cost of preventing or coping with security breaches must fit the probable impacts resulting from those security breaches. Within the IA framework, one of the methodologies that has emerged is a Risk Management Methodology, consisting of an assessment of risk, vulnerabilities, and threats.

Risk	• Understanding exposure to threats • Assessing likelihood of attack and success • Performing up-front and on-going risk assessments that attempt to quantify likelihood and cost of a breach
Threats	• Understanding source and means of particular types of attack • Threat assessments are performed to determine best method(s) of defense • Organizations perform penetration testing to assess threat profiles
Vulnerabilities	• Weaknesses or flaws in a system that permit successful attacks • Can be policy-related as well as technology-related • Vulnerability assessment should be performed on an on-going basis

Table 1. Risk Management Methodology.

Information assurance is implementation-driven

IA solutions implemented in LTE networks for public safety ultimately must be implementation-specific, driven by the unique requirements for security of all of the functions within an LTE network for public safety. IA best practices specifically balance the need for security with the need to impose as few additional operating requirements on users of the LTE network as possible.

Another key benefit of IA for LTE is the application of a minimum number and type of security requirements to ensure interoperability, without limiting the ability of specific jurisdictions, or a future nationwide governing entity, to go beyond these minimum requirements. In typical corporate networks, security requirements consist of commonly held IT best practices where the focus is on connectivity. However, on an LTE network for public safety, there is arguably more at stake. The complexity of stakeholders, systems, devices, networks, and environments precludes just the baseline standard IT security techniques implemented as a one-size-fits-all security solution. Therefore, the additional criteria outlined with information assurance must be used to select the measures appropriate for the agency.

By implementing IA best practices within the LTE network, security benefits are available now, and the network will meet FCC requirements for cyber security and critical infrastructure survivability. Further, the LTE network is designed to meet the evolving technical framework of the FCC's Emergency Response Interoperability Center (ERIC), which is required for eventual integration into the planned National Public Safety Broadband Network. Finally, these additional criteria must take into account the constraints posed by device and network technologies, legacy systems, organizational structures, compliance mandates, regulatory and legal policies, and cost criteria.

Conclusion

In summary, LTE public safety communications systems are critical networks and are essential to the welfare of the country. These important assets must be protected with the same rigor as any other critical infrastructure. Malicious cyber attacks on our communication networks have increased dramatically, and these attacks continue to mature in scope/scale, complexity, and sophistication. The need for increased cyber vigilance has never been greater. Enhancing network connectivity and interoperability to both private and public networks improves information sharing and increases situational awareness, but it also elevates the vulnerability of these networks to externally mounted attacks.

LTE networks for public safety must be strongly safeguarded and proactively monitored completely—from end to end—in order to avert casual as well as advanced persistent threats. The recommended solution is to implement a comprehensive and well-defined security architecture that is organized and managed on a local, or even national, footprint by a trusted, experienced security engineering resource. Further, this must be as capable as those relied upon by government agencies such as DHS and the DoD.

GREG HARRIS is the Cyber and Information Assurance Solutions product manager for Harris Public Safety and Professional Communications business. Prior to joining Harris, Greg was with the Military Medical Command (MEDCOM), assigned as Information Assurance and Governance Program Manager for the Southeast Regional Medical Command, Department of the Army. Greg has more than 15 years of information technology experience, with emphasis on network security, information assurance, and IP networking. He holds a Bachelor of Science in Business Management from Troy University and industry-specific certifications such as CIO/G6 CISSP, CompTIA Security + 2008, and Certified Associate in Project Management (CAPM).

ENDNOTES

1. Nicholas Falliere, Liam O Murchu, and Eric Chien. *"W32.Stuxnet Dossier"*. Symantec, February 2011, Version 1.4.

2. Identity Theft Resource Center, "ITRC Breach Report 2010 Final", December 2010.

3. Grimes, Roger A. *"The true extent of insider security threats"*. InfoWorld, May 11, 2011.

4. D. Martinez et al, "Emergency Response Interoperability Center, Public Safety Advisory Committee (PSAC), Considerations and Recommendations for Security and Authentication Security and Authentication Subcommittee Report", May 2011.

5. D. Martinez et al, "Emergency Response Interoperability Center, Public Safety Advisory Committee (PSAC), Considerations and Recommendations for Security and Authentication Security and Authentication Subcommittee Report", May 2011.

Chapter 16: Apps for Government on the Nationwide Public Safety Wireless Broadband Network

BY BILL SCHRIER

In February 2012, Congress authorized the construction of a Nationwide Public Safety wireless Broadband Network (NPSBN) and allocated $7 billion over the next 10 years to create the network.[1] The NPSBN will be a 4th generation (4G) wireless network, using a standard commercial technology called LTE (long-term-evolution). Hundreds of telecommunications carriers worldwide are planning LTE networks. The NPSBN will be unique because it will be "hardened" against natural disasters and use spectrum dedicated for public safety use so consumer uses (such as teenagers watching MTV on their smart phones) will not reduce its capacity.

Most importantly, Congress essentially broadened the definition of public safety, recognizing that electric and water utilities, transportation and transit departments, and many other government services have a "public safety" function, especially in disasters. Even privately owned ambulance services and utilities and certain other commercial uses will be allowed on the network.[2]

No matter how the Nationwide Public Safety wireless Broadband Network is constructed and operated, it will fundamentally change the way government delivers services—not just public safety, but all critical services, even those provided by private companies (e.g., solid waste or electric utilities).

The recent history of consumer devices and networks is fascinating. Before 2007, few people envisioned smart phones, tablet computers, "downloading apps" or watching TV on a mobile phone. Now, more than 1 million apps exist, and both the iPhone and iPad have taken the world (or at least the United States) by storm. Windows 8, due in late 2012, will cause a similar explosion of use and applications, especially in businesses and government. All these devices, of course, use commercial networks constructed by Verizon, AT&T, U. S. Cellular, T-Mobile, Sprint and other telecommunications networks.

What will happen when the NPSBN is available?

Undoubtedly, as the network is built over the next few years, there will be an eruption of new applications from all sources. "Apps stores" for public safety and general government will spring up. Enterprising firefighters, police officers, and other government and technology workers will write a host of new apps to help them do their daily business. "Communities of Practice" (think "Facebook for cops" or building inspectors) will spring up.

And government services—especially public safety services—will change forever and (for the most part) for the better. What are some of these apps, and how will they change the way government serves citizens?

Access to Computer-aided Dispatch (CAD), Records Management (RMS), and Work Management Systems (WMS)

Almost every public safety department uses commercial, off-the-shelf (COTS) software to record calls to 911 for service and to dispatch those calls to officers in the field via computer-aided dispatch (CAD). In non-public safety agencies, such systems are often called work management systems, but the function is the same: to create and store records of work to be performed and dispatched to individuals and crews in water, electric, gas and street utilities, building and land use departments, and so forth.

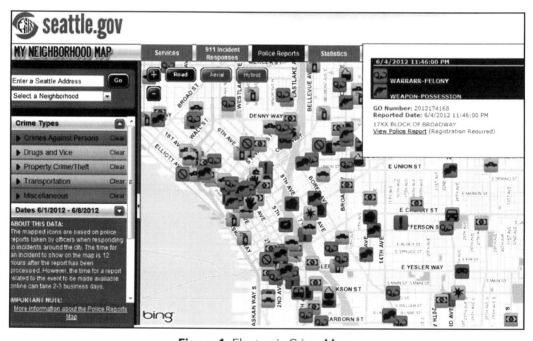

Figure 1. Electronic Crime Map.

In addition, public safety departments usually have a records management system (RMS) to store reports, images, and related materials about crimes, medical incidents, fires, and similar reports. Other government services will have either an RMS or a content/document management system.

Apps will be created to access such systems from smart phones and tablet computers. Such apps will allow police officers to take a crime report in the field using a tablet. The officer will record images and video, victims' statements and even field interviews of witnesses and suspects. Perhaps the new Miranda warning will be, "You have the right to remain silent. Anything you say to this iPad can and will be used against you in a court of law." These devices might even collect evidence, lift fingerprints or DNA samples, and then upload almost all the material via the network.

Police officers will be able to immediately search the RMS for similar crimes and even apply business intelligence to look for similar *modus operandi* for the crime. Such business intelligence applications also will help officers in the field best determine how to patrol neighborhoods, and might even "predict" where crimes will occur.[3]

Similar apps for dispatch and records management will transform emergency medical care, the fire service, utilities, and other services.

Building permits and field inspections

Using the NPSBN, building inspectors in the field would have immediate wireless access to the complete file of a building permit, including any plans, notes, and previous inspections. Using a tablet or smart phone, the inspector could take images and photos of the construction site, make annotations on plans or the images, and immediately file the inspection report both with the contractor and the building permit system.

Building inspectors perform two critical public safety functions. First of all, by ensuring compliance with codes, they protect the health and safety of homeowners and anyone who lives or works in a building. Second, after a widespread disaster—earthquake, tornado, hurricane—they certify whether buildings are suitable for habitation and work, or need repair and demolition. Such functions are entirely appropriate for a public safety wireless network.

Indeed, the entire industry of designing and constructing buildings will undergo a transformation over the next few years. "Blueprints" will become "smart documents" rendered on a slate or tablet computers. Architects will build the blueprints electronically and will specify all the components of a building, right down to the doorknobs, in the electronic plan. Each of the building elements will contain inherent metadata and links to outside specifications. Contractors will take this electronic plan, send it to subcontractors, and quickly generate bids. Field superintendents will use the electronic plan when constructing the building, right down to ordering the individual components and making sure they arrive on site exactly when needed. In fact, if a component such as a doorknob goes out of stock or is discontinued, everyone from the architect to the supplier to the contractor can rebid the project with appropriate substitutes within minutes.

Government building and land use departments will need to use the same electronic plans and specifications when conducting inspections.

Facial recognition

The FBI is building facial recognition apps. Using such apps, police or probation officers will take a photograph of a person's face or an identifying mark (scar, tattoo), and search a database to positively identify suspects in the field. Such apps certainly are imperfect, but hold great promise to positively identify criminals and wanted people. Such apps might be invaluable in missing persons cases, child abductions, and AMBER alerts.

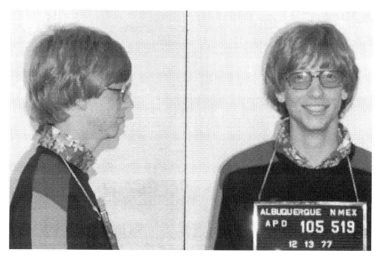

Figure 2. Facial Recognition.

Both an interesting and frightening (for the social and legal implications) application is connecting facial recognition to traffic cameras or video surveillance cameras. Such applications might allow rapid identification of criminals and terrorists in crowds or while driving a car. Such use presents significant potential for violation of individual rights and privacy, however.

Tagging of faces is now commonplace in Facebook, YouTube, and similar social media applications, which (for better or worse) represents yet another source for such identifications. In fact, the dangers to privacy may be greater from such commercial companies than from governments, which, at least, are controlled by elected officials and laws.

Real-time video

Video surveillance cameras are widely deployed in both public and private places, on the street, in businesses and homes. Increasingly, such cameras are networked, either to private networks or to the Internet. With the NPSBN, selected video streams will be broadcast directly to smart phones and tablets in the hands of officers in the field.

Such apps would allow police responding to a bank alarm, for example, to actually view video from cameras inside the bank. Firefighters responding to a hazardous materials spill could see inside buildings or the area of the potential spill. With sensors and "sniffers" mounted with cameras, firefighters and others might be able to determine the toxicity of the spill as well as its extent. Emergency medical technicians, triaging and treating patients in the field, will send real-time video of the work to physicians in hospital emergency rooms for consultation and advice.

With the increasing use of webcams in homes for security and other purposes, officers responding to domestic violence calls or fires in homes might even—with the permission of the caller to the 911 dispatcher—view video and determine the best course of action before arriving on scene.

Many legal implications abound in use of such video, for example chain of evidence and reliability of the video. Could taped video sent to a 911 center from a home with permission of one spouse, for example, be used in criminal or civil trials for domestic violence?

Next Generation 911/311

Many police and fire dispatching centers are installing technology to support Next Generation 911 (NG911). Today, when you need a cop or emergency medical treatment, you "call" 911—"call" as in voice telephone call. Yet, more than half of all phone purchases are now smart phones, capable of sending and receiving video, photos, and other images.[4] Even older "dumb" mobile phones support text messaging. Yet, almost all 911 services support only voice telephone calls—not even text messaging. NG911 is a nationwide initiative[5] which will allow public safety answering points (PSAPs) to accept text messages, photos, and video from callers. These callers will, of course, use commercial cellular telephone networks to make their calls.

But how will dispatchers, in turn, send such material to officers in the field? For example, a caller may witness a robbery in progress at a convenience store and snap a mobile phone photo of the criminal. Such a photo will be extremely useful to responding officers, and with NG911, the photo could be sent to the dispatch center. With the NPSBN, such material can be dispatched over a secure channel to responding officers as they are speeding to the scene.

Figure 3. NG 911.

Similar functionality could be incorporated into 311 systems as well for a wide variety of government services, ranging from hazards (street potholes, downed electrical lines) to simple requests (graffiti clean-up). Again, using the NPSBN, images and video could be dispatched to any government worker in the field, both securely and quickly.

Building and school diagrams, video

Using building plans, firefighters could tag the location of hazardous materials and hose connections in a building. Police officers could see the location of video cameras on a diagram, and, using the 4G network, actually view the cameras inside a commercial building or school in real time as they are responding. Such streamed video would be especially useful in school or shopping mall[6] situations, such as lockdowns with a potential shooter in the school building or mall. Responding officers might be able to quickly determine the location of the shooter(s) and the situation by viewing video from cameras inside the buildings.

Schools also could be protected from sexual predators, as video cameras around schools along with facial recognition might quickly identify known sex offenders and drug pushers in the vicinity of school buildings.

License plate and vehicle recognition

Many cities have now implemented license plate recognition (LPR) systems for routine policing duties such as stolen vehicle location and parking enforcement. Police vehicles specially equipped with LPR video cameras drive the streets snapping photos of vehicles and their plates, converting the plate to text, and then using wireless networks to access remote databases of stolen vehicles.

Parking enforcement officers also use LPR to look for stolen vehicles and to look for parking scofflaws—vehicles with many parking tickets or other violations.[7] Such vehicles can be booted or towed, freeing parking spaces as well as enforcing parking ordinances.[8]

LPR also is used to enforce "two-hour parking" and similar rules. Parking enforcement officers use LPR to record license plate numbers of all cars parked on a particular street, then drive past again in two hours and ticket violators. Most businesses desire such enforcement to make sure parking is available for customers.

Both parking enforcement and LPR require high-speed wireless access to remote databases. Such systems presently use commercial networks but will be enhanced by the speed and security of the NPSBN.

Video traffic cameras are already ubiquitous in many cities. The City of Seattle has more than 160 such cameras deployed, most of which are viewable by anyone who connects to the city's website.[9]

In 2012, full-motion video was added to the public website for a number of those cameras.

Some jurisdictions—and specifically Seattle—have connected traffic cameras to license plate recognition (LPR). LPR recognizes plates as vehicles pass at two points along a route, then calculates real-time travel times between those points in the city. These travel times are displayed on the website and also on readerboards above selected streets. Such timings give motorists an expectation of how rapidly traffic is moving.

Such systems could further be connected to criminal and stolen property databases, allowing identification of stolen cars, wanted criminals, abducted children, and other criminals using cars for crime. Theoretically, the license plate could be compared to the make/model/color of the car to find out-of-date registrations or similar crimes.

Video cameras even might be connected to facial recognition databases, although the quality of cameras, reflections from windows and similar problems exist to prevent widespread facial recognition today.

Such systems pose many difficult and frightening legal and privacy issues. Used without care and proper safeguards, they might spell an end to anonymity in travel on public streets.

Smart grid

Electric, water, gas, and wastewater utilities manage their networks using SCADA[10] systems.

Some electric, water, and gas utilities have installed wireless meters on homes and businesses. Such wireless metering allows not just frequent (say, every 15 minutes) meter reading, but also allows network controls such as near-instantaneous turning services off or on at a premise, or rapidly determining the source and effect of a power outage in a neighborhood.

Smart grid is more than just meter management. Wireless control and management points could include the many transformers, relays, substations, and other places where an electric grid is managed.

Wireless meter control and management is not bandwidth intensive—a few bytes of data for each meter read. Using the many priority-setting capabilities of LTE, such reading can be placed at the very lowest priority of network use, and even shut down when the network is overloaded during a major public safety incident. But such use brings enormous public safety value; the NPSBN will be highly secure and protected from cyber threats—just what electrical utilities need. Electrical network outages are a major public safety threat—by using smart grid, the source of such outages will be quickly determined, resulting in faster restore times and improved public safety. Electric utilities will be able to quickly determine which critical buildings are off power (e.g., hospitals, nursing homes) and prioritize them for restoration.

And such use allows individual homeowners and the utility to see and manage their electric use much better than traditional once-a-month reads. Homeowners can watch for spikes in their electricity use and take appropriate action to reduce such use.

Most of the previous discussion also applies to water, gas, and wastewater networks, all of which are critical to the health and safety of communities, and all of which can be managed as a grid as well as by individual meters.

Home and business SCADA

Just as large metropolitan and regional utility grids are controlled by SCADA systems, homes and businesses have electrical and water "systems" or networks inside their buildings. The "network" elements in these internal grids include electrical outlets, appliances, water heaters, shower heads, faucets, gas appliances, lights, lamps, washers and dryers, and so forth. In the future, each of these "elements" will be monitored and managed via mini-meters. In this fashion, a homeowner could immediately see how much electricity the dishwasher takes to wash a load, how much gas a dryer uses to dry a load of laundry, and even how much water a teenager uses when taking a shower. With the appropriate software, a homeowner or business owner easily can monitor energy and water use, and take appropriate steps to reduce or manage such use. This, in turn, should result in better conservation of these resources.

This application does not have a direct need to use the NPSBN—most of the data could be collected and used by a computer in the household, or sent over the Internet back to the utility for analysis. But the application is listed here because it directly relates to the use of the NPSBN for smart grid.

Parking management

On-street parking is poised to undergo a second digital revolution. The first revolution moved parking management from mechanical meters and manual collection of coins to parking pay stations, where kiosks accept cash and credit cards and spit back sticky pieces of paper that are placed on cars in parking spaces.

The second revolution will allow management of individual spaces. San Francisco,[11] Los Angeles and a few other cities are experimenting with systems where a small wireless device is placed in the street in each parking space. The device tells a parking management system whether or not a space is occupied. This information, in turn, is available to drivers who can determine where the most open spaces are at, which reduces continuous "driving around the block" to find such spaces or wait for them to open.

Many further improvements are possible and under pilots. "Pay by phone" systems, such as the one in Vancouver, British Columbia, allow drivers to pay with their phones to extend the time on their "meter." Perhaps drivers could look for spaces, send a signal from a phone, tablet or computer to reserve them, and a little yellow flag would go up on the meter preventing others from parking in the spot until the payee arrives. Such systems could be extended to private parking lots and garages. Entire maps or mashups will be built showing parking availability and how it ebbs and flows in a city.

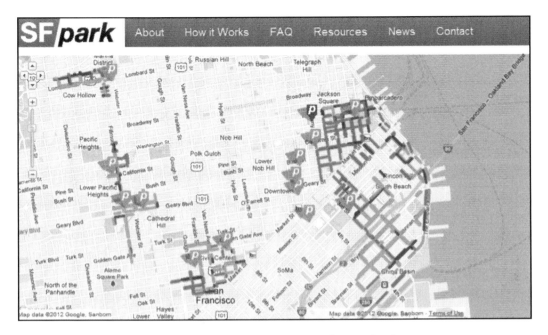

Figure 4. San Francisco On-Street Parking—sfpark.org.

Again, the NPSBN would not be used for the consumer applications of such a system, but could be used for wireless transmission of empty space information, and to better secure the wireless transmission of credit card and payment information from kiosks to payment authorization servers.

Smart cars, drones, and robotics

Another potential app for the NPSBN involves the remote control of certain vehicles, robots, and other equipment. The most immediate application is for vehicles used in bomb squads, hazardous materials spills, and other robotic vehicles. These robots allow investigation, management, and cleanup or defusing of such incidents by remote control. Most of such apparatus would also carry video cameras and sensors. All the video streams, telemetry, and control of the devices might be accomplished via the NPSBN from hundreds or thousands of feet away from the incident site.

Remote-controlled pilotless aerial planes ("drones") are extensively used in the war on terror, in Afghanistan, and other theaters of operations. Such drones are also being deployed in the United States for public safety and potentially firefighting. While the actual control/piloting of the drones might be accomplished through the secure NPSBN or through other means, at the very least, all the telemetry and video could be transmitted via the NPSBN, especially to units and vehicles around an incident scene on the ground.

Even larger vehicles, such as police cars or firefighting equipment, could be remotely controlled, monitored, and managed through the network.

Personal, vehicle, and incident area networks

A potential limitation of the NPSBN is its reliance on cellular technologies. In an LTE network, a typical cell site has three sectors aimed in three different directions, and each sector allows 60 to 200 individual users to connect and use the site. These numbers are sufficient for normal daily activities.

But small disasters and even daily incidents are quite concentrated geographically. A protest demonstration that becomes violent might involve dozens or hundreds of police officers, plus supporting officers from emergency medical, transportation, and other government services. Live video and other bandwidth-intensive applications are often used to manage such a situation. Such an incident might exceed the capacity of a single LTE sector.

LTE has a number of mechanisms to help manage such a situation, such as prioritizing users and applications. Personal, vehicle, and incident area networks will be additional tools to extend the capabilities of the NPSBN.

A personal area network would surround a single officer in the field. A firefighter, for example, might have a body-worn video camera and audio pick-up to record and stream her work on a hazardous material spill. Sensors will also monitor the firefighter's physical condition (blood pressure, temperature, heart rate, and more). And the firefighter may need other wireless devices—smart phone, tablet computer, voice radio—to do the work. All of these wireless connections might be streamed through a single device, and the personal area network will allow the devices to work together and stream data to and from remote locations, such as incident command, 911 dispatch, emergency operations centers, and hospitals.

Vehicle area networks will work in a similar fashion, with a single connection from the vehicle to the NPSBN. The vehicle area network, using Wi-Fi or a similar technology, might connect a dozen or more devices in use around the vehicle.

Incident area networks work on a slightly larger scale, where one or more LTE "cell sites on wheels" will be deployed at the incident to interconnect all the officers, vehicles, and devices in use at the incident, providing additional capacity, as well as connection to the NPSBN and locations, applications, and databases elsewhere in the city or across the nation.

Smart cities

A number of researchers and private companies are spending considerable time and effort to imagine and build components of "smart cities."[12] Such components include many of the ones listed above, such as smart grid and personal or vehicle area networks. They also include a number of useful, interesting, and vital applications for consumers and businesses,[13] such as:

- Two-way and multi-way video conferencing and interactions from mobile devices;

- Heads-up displays, using facial recognition connected to social media (Facebook, Flikr, Pinterest), so people walking down the street wearing "smart glasses" would quickly recognize friends on the street and have context for interacting;

- Building and object recognition, in context. A tablet or smartphone photo of a building will connect to remote databases to search history, businesses located there, coupons, and even building diagrams and internal maps;

- Real-time transit and bus route information and schedules, obtained simply by pointing a smart phone camera at a bus or bus stop; and

- Remote control of appliances, heating, air conditioning, video cameras, and other electronics at home from a smart phone or tablet.

Many of these applications also will be useful for public safety officers and government officials who protect and render services to citizens and businesses. With the Nationwide interoperable Public Safety wireless Broadband Network, such applications will be securely available to handle both daily incidents and needs, as well as during disasters.

Summary

Such applications are, really, only the "tip of the iceberg." As we've seen with the introduction of the smart phone and the tablet computer, thousands—perhaps tens of thousands—of such apps will be developed by private companies as well as enterprising firefighters, public employees, and other apps developers. Just as with the original IBM personal computer or the Apple iPhone, we cannot completely envision today what applications will be useful in protecting and serving the citizens of our cities, counties, states, and the nation in the future. Once the platforms and network are available, innovators will capitalize rapidly on the opportunities.

Bill Schrier served the city of Seattle for nearly three decades, with the last third of that as head of IT and CTO/CIO. Having retired from the city, but not from the IT arena, Bill is now the Deputy Director of the Center for Digital Government, owned and operated by e.Republic. Bill was elected as a PTI Fellow in 2010 and was also elected as a Fellow of the Association of Public Safety Communication Officers. Schrier has been very active over the past few years with public safety communication systems, and in particular the designated national "D" Block. In his earlier career, Bill was a police officer in the City of Dubuque, Iowa.

ENDNOTES

1. How the NPSBN will be constructed and operated is described in a separate chapter elsewhere in this book.

2. The law specifies that first responders—law enforcement, firefighting, and emergency medical services—will have priority over all other users of the NPSBN.

3. Los Angeles and other police departments are already trying "predictive policing:" http://www.slate.com/articles/technology/technology/2012/06/nicole_okrzesik_juliana_mensch_could_cops_use_google_to_prevent_murder_.html.

4. Neilsen: http://blog.nielsen.com/nielsenwire/online_mobile/smartphones-account-for-half-of-all-mobile-phones-dominate-new-phone-purchases-in-the-us.

5. United States Department of Transportation: http://www.its.dot.gov/ng911/.

6. Examples include the Columbine High School shooting and the Tacoma Mall shooting of 2005: http://en.wikipedia.org/wiki/Tacoma_Mall_shooting.

7. Seattle's Parking Scofflaw ordinance is here: http://www.seattle.gov/scofflaw/faq.htm.

8. In Seattle, 900 cars were "booted" and $1.2 million in unpaid fines collected in the first three months of an LPR-enforced program: http://www.seattlepi.com/local/article/900-Seattle-parking-scofflaws-got-boot-this-year-2406326.php.

9. http://web5.seattle.gov/travelers/.

10. SCADA is system control and data acquisition. SCADA systems include electronic monitors and controllers at many critical network control elements to help utilities monitor and manage their networks. Such network control elements include meters, transformers, generators, substations, water/gas control valves, etc.

11. San Francisco's application—both web-based and mobile—is one of the best: http://sfpark.org.

12. See, for example, a series published in the online magazine *Urban Times* here: http://www.thcurbn.com/category/feature/future-of-the-city/.

13. A short, intriguing, video demonstrating these capabilities is at http://www.theurbn.com/2012/06/the-future-of-the-city-8-sketches-of-the-meta-city-video/.

Chapter 17: Critical Operations & Infrastructure

BY *JEREMY SMITH, LORI KLECKNER, SHERRI POWELL, LOUISA KING, SHARON COUNTERMAN*

Technology leaders in the public safety arena are being presented with an overload of information regarding the next generation of 9-1-1 services. It often can be difficult to sort through the information and determine what is needed to begin moving toward a Next Generation 9-1-1 (NG9-1-1) network; as such, data centers, cloud computing, and network and cyber security are discussed below.

Next Generation 9-1-1

In the past 15 years, there have been dramatic changes in the telecommunications industry. To keep pace, public safety and emergency communications systems have had to adapt; however, the aging systems that public safety has relied on can no longer keep pace with industry changes.

The traditional wireline home phone continues to be replaced by Voice over Internet Protocol (VoIP) phones and wireless cellular phones. 9-1-1 calls from wireless devices continue to increase. In 2010, Federal Communications Commission (FCC) Chairman Julius Genachowski stated that almost 70 percent of calls made to 9 1-1 were from mobile phones.[1] To deliver 9 1 1 calls from new services, a work-around had to be developed, which added expense and created additional steps for public safety answering points (PSAPs).

In the 1990s, texting was added to wireless phones. This technology rapidly became the communication method of choice for the hearing impaired and the younger population. Current 9-1-1 networks cannot deliver text messages (calls) to PSAPs. Recent incidents, including the 2007 Virginia Tech shootings where students tried to text 9-1-1 for help, underscore the technology limitations of the current 9-1-1 system and the need for a more robust and flexible 9 1 1 communications system.

The central theme throughout all major visions of the next generation is an IP-enabled broadband network that can share voice, video, and data. This network must be a mission-critical, secure, and fail-safe public safety system.

NG9-1-1 systems operate through Emergency Service IP Networks (ESInets) and are designed to dramatically shorten the time required to identify a caller's exact location. As opposed to current 9-1-1 networks that route a call to a PSAP and then query a database to determine location, NG9-1-1 attaches addresses or coordinates (longitude/latitude or

X/Y) at the beginning of the call process. NG9-1-1 provides enhanced interoperability, ease of use, resiliency, and the potential to spread call workload and costs among parties that share the infrastructure.

NG9-1-1 has been described as a "Network of Networks." NG9-1-1 will permit calls to be transferred easily within networks and among regional networks (ESInets) in a state. The bandwidth of the network is, for all practical purposes, unlimited. It can be shared among all services, allowing interoperable communications and converged applications for all public safety departments in regions that are part of the IP backbone. When agencies converge many applications into a single communications link, potential points of failure are minimized, and duplication of technology is reduced.

NG9-1-1 will provide the following enhancements:

- More accurate location information delivered to call takers

- Improved first response based on crash data delivered from telematics providers to call takers

- Better access to 9-1-1 through text or video chat with 9-1-1 for persons with disabilities

- Ability to re-route calls quickly in the event of a disaster

- Fail-safe and redundant system meeting the needs of public safety now and scalable to meet future needs

The figure below compares the capabilities of the 9-1-1 system in today's environment with the 9-1-1 system in the next generation environment.

Today's 9-1-1	Next Generation 9-1-1
Virtually all calls are voice callers via telephones over analog lines.	Voice, text, or video information, from many types of communication devices, sent over IP networks
Most information transferred via voice	Advanced data sharing is automatically performed
Callers routed through legacy selective routers, limited forwarding/backup ability	Physical location of PSAP becomes immaterial, callers routed automatically based on geographic location, enhanced backup abilities
Limited ability to andle overflow situations, callers could receive a busy signal	PSAPs able to control call congestion treatment, including dynamically rerouting callers

Figure 1. US DOT: Comparing 9-1-1 to NG9-1-1.

Getting started

Whatever stage of development your NG9-1-1 network is in today, it must assume 10 specific roles in a comprehensive approach. From a technology standpoint, NG9-1-1 demands an integrated approach that incorporates four key elements:

- an IP-enabled broadband network,

- data to store and manage location information in new, more effective ways,

- applications that enable both traditional and advanced multi-media 9-1-1 services, and

- a foundation of cyber security hardware and software measures to protect the network from being "checkmated" by malicious opponents or accidentally taken down by users.

Six operational components encompass the technology:

- **Operations, training, policies, and procedures**—This includes how the network is to be used and not used, how applications are to be introduced and rolled out, and how the network will comply with government regulations.

- **Governance**—This includes procedures agreed upon by representatives of all stakeholders sharing the network.

- **Regulatory and legislative**—This includes aligning operations with state regulations, which may need to be revised to accommodate NG9-1-1.

- **Planning**—This is basically a roadmap for strategically planning and transitioning to NG9-1-1 systems.

- **Maintenance, management, and support**—This includes managed services that may be required for selecting and evaluating software, filling staffing needs, maintaining network security, and other purposes.

- **Funding**—This entails a funding plan and source to encompass all entities that may access the network; funding legislation may need to be changed to account for new technologies accessing the network.

These "pieces" may require coordinated efforts with telecommunications experts.

Questions to ask about IP networks

Among the questions agencies should consider when planning the installation of an IP-enabled network are:

- Who will install and maintain the IP lines?

- Who will host the system? Should it be co-located at a carrier's facility to be managed?

- How will the network be configured? Who will manage the firewalls and routers—existing staff or outsourced experts?

- What functions will be placed on the IP network, and what will be the resulting demands for bandwidth allocations? Will mapping or radio integration be integrated into the network before the region goes to full NG9-1-1?

- Who should share a full NG9-1-1 network?

Answers to these questions will vary.

Data centers

With the growing economical need to combine resources, facilities such as Emergency Call Centers are becoming data centers. A data center is often defined as a facility used to house computer systems and their associated components. A data center offers resiliency in the form of redundancy and diversity. For our purposes, the data center holds the technology that supports public safety in the form of telecommunications, networks, servers, databases, and workstations.

Why?

With the need for near zero down time in emergency communications, the benefits of a quality data center facility are rapidly becoming a necessity. When considering a data center, features should be scrutinized. If you are not familiar with the specifications of a data center, keep in mind that retrofitting a building is generally cost prohibitive. When choosing a data center design, flexibility and scalability should be considered.

Physical aspects

The building itself has specifications to be considered. How secure and protected is the building from natural disasters or man-made disasters? The building should support the purpose intended, with raised flooring and adequate space for expansion.

Below are a few general points to consider, which should not be considered an exhaustive list.

- Location
  - Out of the flood zone
  - Quiet area (low traffic, low crime)
  - Easily protected

- Building selection
  - Construction material (internal and external)
  - Floor plan (layout)
  - Easily secured

- Diverse connectivity
  - Power
  - Network

Location

When considering the location to house equipment, economies of scale are realized with regard to providing the ideal environment and consistency across equipment. High levels of conformity and failover processes often are difficult to provide at smaller, separate locations with a variety of specifications. A centralized operating location in the form of a data center offers:

- Hub for telecommunications

- Appropriate environmental controls

- Fire suppression

- Emergency backup power

- Increased security

- Ease of device management

- Ease of monitoring

An example of equipment redundancy for the network is that data centers have spare equipment both online and additional spares. If a router goes down, the online spare takes over without notice, the faulty equipment is immediately replaced, and a new spare quickly arrives on site. This also sets the stage for data consistency and availability.

Building selection

Building materials should have the ability to survive natural disasters (earthquakes, hurricanes, tornadoes, and heavy rain [mud slides, flooding]) depending on the geographical area. Material should support maintaining a proper building temperature and humidity, as well as providing a barrier for electromagnetic fields, both entering and emanating from the building. If windows are present, they should support security (opaque) and sustainability (shatter proof).

Inside, the building should support the weight of equipment placement, provide anti-static properties and not inhibit cooling and humidity levels. Materials should be of appropriate fire and combustibility ratings. Sound barriers and minimized vibration may be required. Keep in mind the placement of fire-extinguishing systems, assuring proper location and types of suppression (water versus chemical).

You should plan for plenty of open space to accommodate growth. Organization of racks and equipment should take into consideration the amount of heat systems produce to avoid hot spots and allow even airflow and temperature. Systems should be organized by functionality, with network equipment separate from servers, and storage systems separate from application processes. It is recommended that staging and testing be located separate from production equipment. Employee workspace should not be co-located with equipment. There should be more than one entrance to the main floor. However, each entrance should be secured with limited access.

The building plan should provide for multiple levels of secured access both within the data center and when entering the facility, such as preventing intruder access via windows or loading docks. Areas should provide security commiserate with content. Highly secure areas should utilize cipher locks as well as biometric authentication systems. Floor and ceiling construction should be taken into consideration to prevent access to secure areas via potential gaps and crawl spaces. Doors should be resistant to forcible entry with secure hinges and proper directional opening. Cipher locks and alarms should be in place at all points of entry to the facility.

Diverse connectivity

Power should be obtained from two separate grids and should enter the building at separate physical locations. Uninterruptible power supply (UPS) and generator power should be adequate as not to interrupt service in the event that both power grids fail. Generally speaking, UPS or online battery power should support systems while the generator is brought online.

Network connectivity should be obtained from two sources that are not dependent on the same physical location or equipment to reach the data center. Check the paths closely to avoid a single point of failure. Again, entrance to the building should be at separate locations, preventing an incident in one area of the building from complete disruption of service. Failover to an alternate network should provide uninterrupted service.

It is important to be detailed and thorough in research and planning. Confirm information such as who provides the power. Data centers utilize a lot of electricity. Can the power company guarantee full power during local peak periods such as the hottest day or when power is difficult to obtain due to a local crisis? Remember, during the most difficult times, you are most likely to be busy, as well. This is no time to lose network connectivity or not have access to your data because the power is out.

As a provided service

You may choose to outsource your data center. You still will need to discuss all the above issues with perspective providers. Plus, you want to know how your data is protected. How is it kept separate from everybody else's data? Who has access to your data? What is their backup schedule? Where is their off-site storage? Is backup electronic and protected? Do they use tapes? How are they transported? What is the recovery time to their disaster recovery site?

Ask about the future. If their customer base grows, can they support growth, or will service suffer? What are their plans for expansion? Do the answers sound logical and well thought out? Another point to consider is your connection to the provided data center. If your data (network traffic) cannot get there, the advantages are moot.

Rating a data center is fairly easy because tiers have already been established. Tiers 1 and 2 do not offer much redundancy and/or diversity, providing a promise of 28 to 22 hours of down time a year. Tier 3- and 4-rated data centers limit downtime at 1.6 to 0.4 hours per year. The key rating difference is the redundancy and diversity provided by a tier 4-rated data center. Tier 4 centers are considered fault tolerant and offer the following:

- planned maintenance that does not disrupt service
- a single unplanned event does not impact critical services
- multiple power paths
- multiple cooling distribution
- annual downtime of 24 minutes

Other points to discuss with a provider:

- Who are their service providers for power and network connectivity?
- Who maintains the equipment and data?
  – Are they certified?
  – Are background checks conducted?
- Separation of data (yours from other customers)

Consider the stability of the company; are they sustainable? Put Service Level Agreements (SLAs) in place, and require the reports to verify levels are being met. Routine reports should include outages, security events, and maintenance schedules. Finally, have an exit strategy built into your contract. If you leave, whatever the circumstances, your data should be returned, and you should have time to transition to a new solution.

Cost

When planning your data center, consider financial feasibility over time. This is either for building your own or contracting as a provided service. Can you financially support your own data center? Will the cost savings be realized over time and fit your financial projections? A tier 4 data center carries a substantial price tag. With raised floors, redundant power, cooling, equipment, and diverse paths to the network, you will pay for the privilege of decreasing your risk of resources being unavailable. Whether you are purchasing a data center in whole or as a provided service, research and careful planning are a must.

When researching your data center, ask a lot of questions and involve people you trust who are well versed in data center design that will be able to provide advice on possible issues. One example of a potential issue: Having two network providers located 500 miles apart is irrelevant if they both use the same local provider to provide service for the dreaded last mile. The last mile is the physical path leading to the building and where each connection enters the building. These paths into the building should be diverse, as well.

Data center summary

Decide why you need a data center and what benefit will it provide for public safety. What is your data center rated? Diversity matters in all areas to assure resources are available 24 hours a day, 365 days a year. Resources include power, environmental controls, equipment, and network connectivity. Your data must be secure physically and in transit.

Finally, what are the costs involved? Whether to wholly own or outsource—does the cost outweigh the risk?

Cloud computing

Cloud computing, sometimes called Software-as-a-Service (SaaS), Service Oriented Architecture (SOA), or Utility Computing, is a computing model where computing resources are provided as a "service" over a high-speed network. Software and hardware (particularly server hardware) is remotely hosted in a data center to which users connect to run their applications. The "cloud" is a metaphor for the networks in which the computing resources reside. Often, virtualization technologies are used in cloud computing models to facilitate scalability, availability, and economies of scale.

Public safety chief information officers (CIOs) are faced with increasing responsibilities, more complex applications, and the need to interoperate—all coupled with less budget and staff. Cloud computing offers CIOs potentially significant benefits, including cost savings, flexibility, and scalability. For example, let's say a region has eight 9-1-1 emergency communications centers. Typically, this would mean that there would be eight full sets of 9-1-1 equipment that may or may not have to talk to each other. However, in a cloud computing model, the server backroom equipment would be hosted at a central location, and only desktop workstations might be left locally. In addition to the obvious savings gained from the elimination of seven sets of backrooms (and all of the ancillary elements necessary to support them), there is a somewhat larger set of equipment at one or two locations. Additionally, the site is able to increase or decrease positions at sites much more easily, and in the event of a catastrophic disaster to one location, calls can be addressed at another location.

However, there are drawbacks to cloud computing: security and privacy, dependence on network connectivity, less direct control, and application support issues. CIOs who consider implementing cloud computing must approach the issue carefully and set practical realistic goals.

Key issues to consider involve:

- **Application support**—Can the applications targeted for the cloud actually live in the cloud? In the public safety market, many proprietary and legacy technologies exist. This may create challenges for moving critical applications into the cloud.

- **Service Level Agreements**—Ensuring that specific and appropriate SLAs are in place is critical. In the event that issues occur, it will be important to have something to fall back on.

- **Private clouds for government**—While commercial organizations may be more willing and able to use public clouds for their applications, a public safety organization may have to consider using a private cloud to deal with security issues. Using a public Internet-like connection may not be palatable to the stakeholders of the system for privacy or security reasons.

Network and cyber security

Public safety systems are migrating from a closed-legacy, analog system to interconnected, IP-enabled, digital systems. Historically, cyber security concerns have focused on physical access to the facilities housing the systems and the four walls protecting them. However, today, technology has advanced beyond closed network systems. The need and/or requirement to interact with our neighbors across networks are far more prevalent than ever before. Public safety systems are attractive targets for hackers or cyber criminals, and the interconnection of these systems reduces the barriers to attack as never before. Public safety CIOs must account for the security of their systems and equipment.

A comprehensive approach to securing your systems is necessary. Several high-level steps should be followed to ensure your systems stay secure:

- **Conduct a threat/vulnerability assessment.** A vulnerability assessment examines and identifies weaknesses or gaps in security. This provides a baseline understanding of the level of security currently in place. The results of the assessment are utilized to create a compliance plan intended to close the gaps or mitigate the weaknesses. Assessments should be performed on an annual basis to track progress, assure changes are properly implemented, and identify new weaknesses that have developed, usually as a result of changes in technology.

- **Create a security plan.** A security plan identifies the goals, objectives, and intentions regarding cyber security in your organization. It is a means to exercise due care by communicating your intentions regarding cyber security, as well as promote and increase awareness of security. It identifies the standards and frameworks applicable by legislative, regulatory, or policy affecting your environment. It is the foundation of your security program.

- **Develop security policies.** Policies are a means to communicate your desired security posture as established by your security plan. Policies speak to specific areas such as passwords and information sharing, addressing expectations for compliance and disciplinary actions that result from non-compliance, and are a necessary part of your cyber security strategy.

- **Implement countermeasures.** Safeguards—such as anti-virus software, firewalls, intrusion detection, and prevention systems—can be implemented to protect against malicious activity. Including security training for employees provides understanding of the do's and don'ts, as well as steps to take when a security infraction is detected.

- **Monitor and maintain.** Security is not a "set it and forget it" activity. Threats change, and so must your security. Recurring monitoring, maintenance, and tuning of your security infrastructure are required to stay one step ahead of the bad guys.

Finally, it's important to note that the public safety arena is, for the most part, somewhat new to the world of cyber security. Accordingly, no official standards and not many best practices exist to help guide CIOs as they seek to secure their systems. However, some guidance in this area to assist states, local governments, vendors, maintenance providers, and content providers in addressing security into IP-connected solutions covers the following areas:

- Security planning and policies

- Information classification and protection

- General security
- Safeguarding information assets
- Physical security guidelines
- Network and remote access security guidelines
- Change control and documentation
- Compliance audits and reviews
- Exception approval and risk acceptance process
- Incident response and planning

Planning is the key. From a security plan to set the processes in motion to planning each step along the way, start with your plan, and build from there. Help is available. Reach out to other CIOs who are willing to share their experience and expertise, and leverage the experience of professional firms who do that for a living.

JEREMY SMITH, LORI KLECKNER, SHERRI POWELL, LOUISA KING, SHARON COUNTERMAN; The L. R. Kimball Team; an iCERT Member Company.

ENDNOTE

1. www.fcc.gov/Daily_Releases/Daily_Business/2010/.../FCC-10-200A2.doc

Chapter 18: A Perfect Storm: Managing Court Technology in the 21st Century

BY TOM CLARKE

Court CIOs face a period of unprecedented change and challenge over the next decade. Courts are innately conservative organizations that have changed little over the last century. They are now suddenly trying to convert all at once from paper to virtual business processes while reinventing almost all of their core workflows, replacing most major applications, and significantly improving overall productivity. Court CIOs are expected to play a key role in all of these revolutionary changes. This paper discusses what it will take for them to be up to the challenge.

A brief history of court technology

Significant use of modern information technology in the courts really started only in the late 1960s. The first case management systems were written in the early to late 1970s. Because none had existed before, each one was a totally new kind of court project, and requirements had to be documented from scratch. Because courts had operated manually before, the bar was very low for project success. The first generation CMSs were primarily systems for clerks to track court cases. There were no court CIOs back then—only project managers.

Like the rest of the world, court IT slowly morphed from early systems based on mainframes and dedicated word processors to client/server systems using mini-computers and personal computers, and finally to web-based systems. Second generation commercial CMSs usually met the high-level functional requirements identified by the National Center for State Courts. The market for commercial CMSs continued to churn as major vendors came and went every five to 10 years. Very small secondary markets in jury management software and electronic filing systems sprang up and found limited business. The jury management market slowly consolidated to two national vendors and a plethora of very small ones. The position of IT manager was called many things and could range in actual responsibility from maintenance of the local LAN and PCs to the director of a large IT shop.

During this period, the underlying business processes of the courts remained largely unchanged. The CMSs automated previously manual data processes and provided little direct support to court administration staff. The same was true for jury and e-filing applications. Courts tended to stay with whatever jury vendor they selected and implemented new CMSs only once every 10 to 20 years. It was not uncommon to see systems even older than that, so court IT staff often were completely inexperienced in managing such large IT projects. A number of IT managers bet their careers on their ability to bring home successful projects, and some of them lost the bet.

It really has been only during the last 10 years that courts have started to undergo a business revolution led by technological change. The rise of the Internet sparked an unprecedented wave of replacement CMS projects with much broader functional requirements. For the first time, the Conference of Chief Justices (CCJ) and the Conference of State Court Administrators (COSCA) officially regarded technology as strategic to courts and mandated support for national technical standards. These national court governance associations also enlarged their official definition of what constituted adequate court technology capabilities (see 2005 COSCA white paper, "The Emergence of E-Everything") and founded an association of court CIOs (Court Information Technology Officer's Consortium or CITOC).

It is during this latest period that the courts started to mimic other organizations by creating true CIO positions. It has taken a full decade for a majority of the state courts to make this change. Predictably, the first courts to recognize the strategic business importance of technology leadership were the big metropolitan courts and state courts with their own significant infrastructure and application development shops. The COSCA "E-Everything" paper was the turning point for many remaining states.

The final impetus for massive change was the Great Recession of 2008. In many states, the cumulative reductions in court budgets went beyond 20% of previous funding levels. Faced with such a deep and prolonged drop in funding, courts were forced for the first time in the modern era to consider reengineering or redesigning their business processes in fundamental ways. For virtually every court, those efforts started with and were built around the use of technology. CIOs suddenly became central figures in the survival of many courts. They found themselves regularly talking to supreme court justices and administrative judges for the first time in their careers.

The problem

Because of the way courts are structured in most states, court governance is a mess at the best of times. Judges are almost always nominally in charge, but judges are lawyers rather than administrators. By definition, they are seldom trained in basic management practices, and by nature, they are inclined toward pursuits of a solitary and analytic nature. When judges are separately elected officials, even the presence of active presiding judges or an assertive supreme court is often not enough to put in place simple governance best practices. There is a well-documented culture of autonomous decision making in most courts that blocks attempts to achieve business process consistency and common policy approaches.

Where a simple top-down approach to governance is not possible, the next best governance model is an executive team consisting of the presiding judge and the court administrator. This combination of roles is more likely to appreciate the importance of technology. An important indicator of the governance attitude toward technology is where the IT director resides on the organizational chart. If she reports to a facilities manager or equivalent position and acts essentially like a line manager on the infrastructure side of the organization, then technology is probably viewed as a relatively unimportant detail of court administration. If she reports directly to the administrator, then technology is probably seen as a strategic enabler of improvements in court administration, and the IT director is likely to behave like a real CIO.

Managing technology projects is a difficult endeavor in the best situations. Many studies document the incidence of failed or partially failed projects. It can be even more challenging when the court managers governing technology projects are uneducated in public administration and often uninterested in the application of technology. After all, most court administrators are hired for their ability to keep judges happy and solve political problems. That kind of skill set does not often include a practical understanding of technology management.

That situation puts a lot of pressure on court CIOs to be successful on their own. It creates a daunting set of constraints on project success for those court CIOs who may have come into their current positions with limited skill sets, experience, and training. In fact, until recently it was not uncommon for court CIOs to rise up through the ranks from lowly beginnings as LAN administrators, desktop support specialists, or even court clerks. It is only fairly recently that a consistently high degree of technical professionalism has started to characterize the typical court CIO. As a result of the current demands in the courts, the position is now undergoing the most rapid evolution in requirements and personnel ever seen in the court field.

Of course, many courts are small and really have no effective IT staff. Instead, the court relies on the city or county IT department, or receives its IT support from the state administrative office of the courts. When it comes to court CIOs, we are mainly talking about the larger metropolitan areas and state court systems. Those larger jurisdictions are also the courts taking on the most challenging IT projects and court reengineering efforts. This group of larger courts splits fairly evenly between those who develop their own applications and those who contract out that capability either to commercial off-the-shelf (COTS) vendors or to true software developers. Those who outsource to commercial providers still essentially are managing the development of custom software, because individual court functional requirements usually are deeply unique.

Courts with their own development shops often lag behind the technology curve for internal technical skill sets. COBOL, Natural, Adabase, and AS400s still are found frequently in court data centers long after the commercial sector has moved on to newer products. Because governments typically underfund training and do so even more during times of crisis, it also has been extremely difficult for court CIOs to upgrade their staff members' skill sets. With wages typically being lower than what the private sector offers, the ability of courts to attract and retain highly skilled staff with new skills or skills in high demand is severely constrained.

Until recently, commercial court software vendors have not done much better at keeping up with best practices for application design. COTS CMS products slowly grew over the last several decades into huge monolithic systems of great tightly coupled complexity. They were expensive to maintain and even more costly to modify as new business requirements emerged every year. Such systems only gradually started to embody improved designs such as modularization, tiered architectures, or web-based interfaces. E-filing vendors also continued to market high proprietary applications long after open national technical standards had been completed. Jury management products operated as standalone islands of functionality. External interfaces to systems in other organizations were expensive exceptions. Only in the last couple of years have commercial vendors started marketing highly configurable CMSs and standards-based e-filing systems. Those same vendors are just starting to take baby steps toward products with public interfaces based on open standards and more modular architectures.

Given the political pressures of the job and chronic underfunding and understaffing, it has been attractive for court CIOs simply to meet the demand for new capabilities by, in effect, turning over their software needs to a CMS vendor, who then acts like a general contractor for court applications. This definitely lowers the risk and management burden for the CIO, but it does so at the cost of control and dependence on a particular vendor. As commercial CIOs evolve into general contractors who manage the integration of many applications and vendors, court CIOs have evolved in the opposite direction toward a role like an operations vice president on the business side.

The job scope

As courts suddenly have been faced with a need to quickly bring about large changes in their business processes using technology, the job scope of the CIO has escalated accordingly. Imagine having to consolidate data centers, carry out server virtualization, upgrade wide area networks, replace major applications, facilitate business process reengineering exercises, and referee competition for scarce technology resources all at the same time. Add to that mix a rapidly changing and partially immature vendor and product market. Finally, stir in a sudden need to implement cloud solutions and other infrastructure innovations in response to a bad recession. That is just on the technical side of the job.

Frequently, the CIO also is being asked to guide major redesigns of core business processes, facilitate discussions among key stakeholders, including judges, about what the new business processes should look like, successfully direct the associated change management initiatives, and do all of these things that scare many users of the systems while maintaining or improving business and political relationships between the IT shop and its court customers. This can create very uneasy working relationships that require highly refined communication skills to manage well. Thus, the CIO must be both master of current technologies and wizard of change management. That is a range of skills not often found in any CIO but nonetheless very often demanded in the current environment.

Required skill sets and capabilities

Courts are recruiting for new skill sets in CIOs as they crave competent technology leadership. The list of desired skills is both broad and long. The raising of the expectations bar is causing some experienced incumbent court CIOs to suddenly find themselves falling short and being replaced or asked to retire. The turnover in court CIOs is now approaching that of the private sector, and it is easy to see why. The most successful court CIOs are simultaneously now technical wizards, governance geniuses, and business process reengineering gurus. That is after they have mastered the core skills of managing budgets, personnel, infrastructure, and projects. Such a broad and demanding job scope is a recipe for disaster unless the CIO is very talented or knows how to surround herself with an excellent staff. Here, then, is a list of the top five skills required for a successful court CIO:

- **Communication Skills, Part 1—Marketing.** There was a time when a court CIO could sit in his cubicle and toil away in seclusion and semi-anonymity. That time is gone. CIOs must lead the charge on change, and that requires superlative communication skills. Successful marketing is the name of the game. Faced with a wide variety of customers and stakeholders, the court CIO must be capable of successfully communicating a vision and business future in many different ways and styles—not least to judges who may have little interest or trained capacity to understand some of the implications.

- **Communication Skills, Part 2—Business Analysis.** Business process reengineering is now a core responsibility. CIOs are expected to lead and manage fundamental redesigns of business processes while still remaining appropriately deferential to the leaders of the courts. That requires both formal training in business process redesign and an ability to create new business processes in partnership with stakeholders. Facilitating the necessary discussion in a non-technical and non-threatening way is a critical skill set.

- **Communication Skills, Part 3—Translation.** Just as in the private world, a court CIO sits on the divide between business people and technical people. She must be able to speak both languages fluently and translate back and forth between the two groups. This is probably the one key skill of the position that is non-negotiable. Without it, a CIO either becomes too technical and loses the support of the policy makers, or too business-oriented and loses the respect of the IT staff. Many a court technology project has gone down in flames due to a "failure to communicate," as Clint Eastwood so famously described it. It is not clear that this skill can be taught, so many courts specifically look to acquire it via the hiring process. Internal IT managers who aspire to become court CIO would do well to demonstrate this capability.

- **Agile Management.** This is the ability to competently outsource much of the court IT capabilities. Many court IT shops cannot develop or maintain their own capabilities either for their infrastructure or their applications. They must assemble best-of-breed business partners, constantly assess whether or not better business solutions are available, and successfully negotiate appropriate contracts and service level agreements—all while protecting judicial independence in the eyes of court leaders. Those CIOs with significant in-house capabilities still need many of the same skill sets. They are just negotiating with staff instead of commercial contractors.

- **Architectural Vision.** The number of technology capabilities now expected by court leaders to routinely exist is large and growing. In the past, a court CIO basically managed the internal office applications, hardware of various sorts, both local and wide area networks, and one very large CMS application. Now, they must manage all of that plus many more significant applications and hardware capabilities, including document management, videoconferencing, digital recording, and more. The CIO does not have to be a technical architect herself, but she does have to recognize the importance of having such an architectural plan and ensure that one is put in place and maintained.

Recruitment strategies

The business pressures recounted above are motivating courts to look outside the courts and the local jurisdiction for CIO talent. A number of recent new state court CIOs have come from the private sector with no court or government experience. A few years ago, that never would have happened. Predictably, their record of success has been mixed. Some have skillfully made the transition from the commercial world to government, and others have struggled to cope with the broad array of stakeholders, diffuse governance, constantly changing business requirements, and limitations on funding and skilled staff.

Another significant new source of court CIOs is the state or local executive branch. These CIOs come with experience in government and often also with a good sense of best practices for technology governance and project management. Because courts tend to lack the tight top down structure of executive branch agencies, these CIOs still find the loosely coupled nature of court organizations a very difficult challenge. Instead of being able to simply tell users to follow orders from their bosses, the CIOs must engage in broad and lengthy marketing campaigns to win over potential users of their systems. Since those users also see what technology can do for them in their private lives, they can be very demanding of CIOs who lack similar resources. Court CIOs from the executive branch must learn to carefully trade their ability to tolerate project risk against the ability to deliver desired new IT capabilities.

Of course, new court CIOs also may come from within the ranks of the hiring court or other courts. The world of courts is a small one, so the reputation of candidates for CIO positions already may be known to the hiring court, whether accurately or not. Perceptions of candidates may prevent some opportunities for advancement and help others. It definitely helps potential court CIOs to become active in CITOC and be known in a positive light by state court administrators. Internal candidates must create the perception with their own management that they can successfully deal with court stakeholders while guiding large technology projects and facilitating business process reengineering efforts. This sometimes can be difficult if they have limited access to senior management or were hampered by the policies of the former CIO.

Courts looking for new CIOs are well-advised to seek a pool of candidates that represents a good cross-section of these recruitment pools. Having candidates from private industry, the executive branch, other courts and the internal IT shop enables the hiring committee to observe the tradeoffs directly and decide what mix of skills are most important to the court. Because a perfect candidate almost must be a superman of sorts, it is useful for the court to be clear about its hiring priorities. That also helps the new CIO understand which job duties will make or break her success.

Additional resources

There has never been more support for court CIOs seeking advice than today. Here are some resources for court CIOs to leverage as they struggle to succeed:

- Project management certifications are rapidly becoming a price of admission for key court technology staff. The organizations that provide such certification are well-known, and training opportunities abound.

- Enterprise architecture is partly a business skill and partly a technical skill. CIOs should have at least a high-level acquaintance with this conceptual area, starting with the excellent products developed by the National Association of State CIOs (NASCIO). NASCIO is the equivalent of CITOC for state executive branch CIOs.

- The National Center for State Courts maintains a technology section on its website where a broad array of information on court technology is available. The content ranges from open court technical standards to blogs on issues of the day and everything in between. Free and contractual technical assistance also is available from NCSC.

- CITOC also offers a very broad range of services from colleagues as the association of court CIOs. It meets once a year and hosts an active listserv, as well.

- The Institute for Justice Information Sharing (IJIS) and the Forum on the Advancement of Court Technology (FACT) are justice and court technology vendor associations, respectively. Many respected business partners of courts are active participants, and these professional associations offer court CIOs the opportunity to become adjunct members and hold conversations with industry outside the normal business context.

- NCSC regularly sponsors two major court technology conferences. On the even years, it is the E-Courts Conference, which focuses on e-filing and related electronic court capabilities. On the odd years, it is the much larger Court Technology Conference, which hosts the largest court vendor show in the United States.

- The Global Justice Information Sharing Committee is a Federal Advisory Committee (FAC) of the Department of Justice that advises the U.S. Attorney General on issues relating to state and local justice information sharing. They maintain a large set of useful products on policy and technical issues at the website maintained by the Office of Justice Programs.

The bottom line

Being a court CIO has never been more challenging, yet it is also a very exciting time. The personal and career rewards for successfully managing court technology today are considerable. Pay ranges are gradually rising as court administrators recognize the critical importance of technology leadership. If you like innovation and risk, being a court CIO is the job for you!

TOM CLARKE has served for the last seven years as the Vice President for Research and Technology at the National Center for State Courts. Before that, Tom worked for 10 years with the Washington State Administrative Office of the Courts, first as the research manager and then as the CIO. As a national court consultant, Tom speaks frequently on topics relating to court effective practices, the redesign of court systems, and the use of technology to solve business problems. Tom currently is working with several state court systems and metropolitan courts on "reengineering" projects to significantly improve their productivity while preserving the quality of the services they provide. He also actively consults on the successful use of technology and best practices surrounding court technology.

Chapter 19: Assuring Change Management Success During Technology Project Deployments

BY RONALD P. TIMMONS

Change is coming to Public Safety Answering Points (PSAPs) in the form of new technologies and expanding expectations from the public. The first generation of centralized 9-1-1 reporting of emergencies relied on voice telephone calls to interface with the public. With the wide proliferation of multimedia wireless devices used by all segments of society, PSAPs of the 21st century are preparing to widen the menu of contact methods available through implementation of Next Generation (NG9-1-1) systems. NG9-1-1 systems will make delivery of video, images, text messages, and other archival data to PSAPs possible for the first time.

An enlightened change management approach will help to address the necessary human factors involved in such a major programmatic change. There is a tendency in technology generation changes to focus on the technical aspects of deploying the hardware and software to the exclusion of human aspects. The systems part is relatively easy compared to getting the people using the technology to accept the change and use it to optimal potential and efficiency.

Managing change is a critical skill for leaders in all segments of the public sector, with several contemporary approaches showing promise toward reversing the long history of failed change efforts. Even if practitioners hope to forestall change and keep everything the same, external influences, such as public expectation to provide enhanced NG9-1-1 service, make change necessary and inevitable.

Applying academic theories and models adds strategic perspective toward improving the chances of success in change implementation. This chapter presents a general overview of change theory, linking the literature into a coherent path of change recommendations for PSAP managers facing difficult organizational change during technology project implementation.

The nature of change in organizations

Despite stoic adherence to traditions, public safety agencies are in a constant state of change, influenced by a mix of leadership styles, prevailing crises, shifting political agendas, and rapid developments in technology. Human beings have a strong penchant for avoiding change, out of fear of loss, blame, embarrassment, or feeling threatened.

Proposing and attempting change are noble pursuits, but having a significant program change successfully become part of the organization's culture is another matter. Establishing lasting change in adults is among the most difficult of all endeavors to achieve.

Record of change program failures

The tendency heretofore has been to force technological change in organizations by management dictates, with predictably dismal result. One study found top-down federal government reduction efforts did not achieve wide success because they failed to consider micro level sensitivities and the ability and tendency of recalcitrant entities to circumvent the change directives (Kettl 1998; Thompson and Fulla 2001). With up to 70 percent of change efforts ending in failure (Higgs and Rowland 2005), something is clearly wrong with the way we have historically approached organizational change. Standard dysfunctional approach to organizational change conjures the often-repeated definition of insanity: doing the same thing repeatedly and expecting a different result. Organizational matters are not always as they appear on the surface. There is inherent complexity among the contradictions and tensions present in the typical workplace.

Human reaction to change

Attention to change management approaches, leadership behaviors, and underlying assumptions of cause-and-effect relationships involving change management leads to better understanding of interdependencies. Abrupt, forced change is seldom sustained. Strategic, careful framing and presentation of a pending program change is a more effective leadership tool than efforts to finesse behaviors (Higgs and Rowland 2005).

Timing is critical for change success; sometimes having the courage to temporarily delay a proposed technological change is the best course of action. The timing and approach of change initiatives transcends any argument or logic supporting the need for the change itself, and even the amount of pressure exerted from stakeholders and executive leadership to make it happen.

It can be difficult to evaluate the net benefit of new technology programs and systems. It is not easy to quantitatively project what would have transpired without the changes. In essence, it is impossible to control for what the status quo would have yielded, once the path away from that reality is deviated. The circumstances would have not stayed the same as the known progression forming the conjecture. Such counterfactuals, or notions of the alternate future being unknowable, are an abstract concept, neither easily explained nor likely acceptable to skeptics.

Factors such as instability, cynicism, multiple, redundant layers of management, and use of innovation for only self-promotion instead of betterment of mission and organization, are especially problematic. The structure of the group may work counter to the long-term success of change programs, regardless of how well-deployed or reasoned any new change effort is. Knowledge of these concerns gives CIOs an opportunity for mitigation, simultaneous to the change deployment considerations.

Public and private sector change approaches

Some ask the inevitable question of whether there is a difference between public and private sector change efficacy and approaches. The format and tendencies of public organizations sometimes work counter to sustaining change, in what Light (1998) calls the "unpreferred states of organizational being."

Robertson and Seneviratne (1995) use a detailed meta-analytic research method to examine the results of change efforts in both sectors. The macro-level findings indicate no significant difference between public and private sector planned change outcomes. Some interesting micro-level differences exist in the area of discretion and control afforded to private managers to make nuanced adjustments as needed, as opposed to the rigid, formatted constraints of traditional non-entrepreneurial public sector management.

One study (Fernandez and Pitts 2007) found that public organizations change frequently, despite impressions to the contrary. Further analysis found that bottom-up forces are in play at public agencies, along with top-down pressures from political entities and other stakeholders. Public sector CIOs work at the vortex of competing pressures for change from those hierarchically positioned both above and below them.

People working together in organizations typically exhibit competitive behaviors, not conducive to the feelings of safety and security needed to make necessary program improvements and course-corrections. Managers can increase the sociability and solidarity of people in the worksite by cultivating an atmosphere of shared interests, reduced formality, workplace caring efforts, and de-emphasis of hierarchical differences, all aimed at improved commitment to organizational goals and success.

Public sector agencies are not renowned for emphasizing informality and having fun at work programs. It is difficult to imagine upbeat and whimsical practices being encouraged at some PSAPs, in part because of public scrutiny of tax dollars spent on anything that resembles fun for public employees, and also because of the deadly serious business at hand in a heavily recorded environment. Despite the inherent limitations, efforts to lighten the atmosphere, within prudent, professional boundaries should be employed to reduce the subtle, negative influences of an overly-negative workplace.

Change models and theories

CIOs stand to benefit from taking systematic approaches to change management. Models bring order and predictability to complex situations, such as the inherently dynamic factors in the typical workplace. In reality, it will probably take the mixture of one or more models to accurately recreate the complex set of interdependencies present in public safety organizations. The use of change models and strategic approaches can be valuable in attempts to "debureaucratize" public safety agencies. The byproduct of the modeling process works to suggest sustainable solutions in the unstable organizational world. Although models provide a somewhat limited view of reality, they do offer the opportunity to examine factors from a systems viewpoint, rather than accepting outcomes as being completely random and unpredictable (Kiel 1989, 1994).

Organizational theorists work to find discernable patterns in what may appear to be chaotic and erratic influences and outcomes. Rather than assuming that the presence of chaos is entirely negative, such periods of chaotic behavior may represent opportunity for organizations to seek positive evolutionary possibilities.

These models offer a glimpse of theories applicable to public safety agency change management:

- Complexity Theory acknowledges a system of interconnected influences present among behaviors and variables within groups of people. The origins of Complexity Theory draw upon scientific notions of even the slightest of environmental factors potentially having significant influences later in the sequence of events. One of the simplest notions of Complexity Theory is that order is a natural byproduct of systems, regardless of the complexity of the system itself (Weick and Quinn 1999).

- Nonequilibrium Theory offers insights into the complex interrelationships within public safety organizations. Borrowing concepts from natural scientists, Nonequilibrium Theory allows for the identification of a phenomenon whereby expected and predicted behaviors can be observed for extended periods of time, interposed with episodes of random fluctuations.

- Dialectical Theory seeks to find harmony, productivity, and stability within a balance of power, following the collision of two opposing concepts.

- Life-cycle Theory shows a common sequence of events—a roadmap of the path organizational events are likely to take—based on common, predictable patterns. There is a pace and rhythm to organizations and systems, not always readily apparent to those embroiled in the middle of the circumstances.

- Teleology Theory provides guidelines for judging the change efforts and the philosophical doctrine guiding the movement.

With so many change management theories published, it is easy to lose track of the main purpose: gaining insight and predictability into complex human organizational behaviors. Van de Ven and Poole (1995) conducted an extensive interdisciplinary literature review of change process theories. Twenty different process theories were identified and grouped into four basic schools of thought.

The four change "motors" are:

- *life cycle* (organic growth metaphor, from inception to termination),
- *teleological* (purposeful cooperation),
- *dialectical* (opposition and conflict), and
- *evolutionary* (competitive survival).

These groupings function to isolate broad categories of change management theories. Readers of the four motors theories are cautioned that organizational change is more complex than any one theory or single motor can articulate, despite the fact that 20 specific theories were condensed to these four major change motors. Interplay between two or more of the four motors needs to be considered when contemplating the complex influences and variables present in public safety organizations. These change motors may operate simultaneously or alternately, producing exponential combination possibilities.

Theories help CIOs see their environments from a philosophical viewpoint, but they are not the whole answer to the complicated change approach dilemma. Despite evidence of their contemplative value, selection of any one specific model is less of a predictor of eventual change success than whether the changes are logical and resonate with the reasons given for the change initiative. Even if CIOs are constrained in their ability to expend resources on specific staff socializing efforts, consistency in stated purpose for the change—combined with visible, tangible needs—will improve change success.

Implementing change

Stakeholder involvement

Although the tendency is for CIOs to rush to the procurement and deployment stages, a period of due-diligence to cultivate stakeholder buy-in increases the likelihood of success. It is tempting to skip the first few steps of the change management process, out of eagerness or ignorance, but the illusion of speed will not sustain and will likely undermine the overall change effort.

Any new initiative should start with building a case for change, invoking appropriate drama, even associating actual or impending crises, as one way to build a case for transformation. The skillful communicator can paint a picture of risk and undesirable consequences if the status quo is allowed to play out, offering the proposed change as the best alternative to plodding into the unknown (Kotter 1995).

In the case of NG9-1-1, CIOs can learn from the Wireless Phase II experience of the last decade. The enhanced location technology was unevenly deployed, opening late-adopters to criticism for missing information during life-and-death situations. Timely deployment of NG9-1-1 assures the public will receive state-of-the-art emergency services. Timeliness in deployment creates moral and practical imperatives, from which a sense of urgency can be based.

A strong sense of need and commitment from midlevel PSAP managers, with their true feelings exposed, is a major priority at this stage. A reasonable target is having three-quarters of the management team convinced that business as usual is completely unacceptable before the change process is initiated.

Typical change management mistakes

Contrasting the typical ineffective approaches with a more enlightened process affords CIOs facing a major change with a sequence of recommended actions:

Typical Dysfunctional Change Management Sequence (Griffith 2002)

1. Big idea by new power broker
2. Inner circle meets—No dissenters
3. Small group rushes changes
4. Limited plan distribution
5. Hasty plan announcement
6. Plan deployed; Antagonists alienated
7. Inner circle closes ranks
8. Minor adjustments made
9. Celebration regardless of sentiments

Eight Steps for Successful Change (Kotter 1995)

1. Establish urgency
2. Form coalition
3. Create vision
4. Communicate vision
5. Empower others

6. Facilitate and celebrate early wins

7. Improve and reinvigorate

8. Institutionalize changes

Managing emotions

Rather than disproportionately emphasizing the cognitive and logical aspects of change, there is value in also acknowledging the emotional elements. Consistent with Complexity Theory, managers need to consider the full realm of elements and influences—big and small—as the change initiative is rolled out. Top-down management styles need abandonment in favor of a "spontaneous self-organizing approach," which allows timely and appropriate responses to environmental changes (McBain 2006).

Times of change call for maximum sensitivity and flexibility on the part of managers. It is interesting to note that middle managers take comfort and assurance when senior executives show empathy toward them, yet they sometimes are slow to nurture the same emotional connection to their own subordinates. One study showed value in acknowledging and validating the human emotions of fear, hope, anger, tiredness, and occasional joy displayed in the workplace (Vince and Broussine 1996).

Change involves moving from a known state to an unknown one, triggering strong emotional responses. CIOs should focus on efforts to acknowledge and minimize the pain felt by those impacted. Allowing organizational members to contribute and be empowered to comment is critical to reducing anxiety and difficulties.

Imagination and rumors may unnecessarily stoke apprehension and passions. Having management emphatically state what is not going to change may prove helpful. Identification of scope and boundaries avoids incorrect assumptions and provides some base of assurance for people to focus upon and from which to take some comfort.

The tendency of keeping emotions out of the change management literature is traceable to tenets assuming emotion is too fluid and intangible for scientific measure (Lewin 1947). Emotions play an important role in successful change efforts, especially relating to workload, uncertainty, management alternatives, and job security (McBain 2006).

Ignoring the emotional elements of the change process is typical, yet emotion management is a significant factor in successful change processes. Freud (1984) establishes that behavior is prompted by both conscious and unconscious mental processes. Situational affect and reactions to workplace situations are in part involuntary and somewhat unapparent to the one experiencing them.

The nature of work is closely tied to notions of self and ego, especially in Western culture, where personal worth is measured closely by one's employment, position, and title. Such close personalization of work, as opposed to international tendencies involving altruistic workplace perspectives consistent with organizational honor and well-being beyond self, makes attention to personal issues especially acute in Western workplaces.

Proper information delivery

By understanding common change concerns, we are more likely to reach the core of personal change objections and anxiety. For large public safety organizations, one approach suggests value in change announcements delivered directly by first-line supervisors, not from slick video presentations or auditorium-style announcements from senior management (Larkin and Larkin 1996). Gathering employees into a large crowd, just as emotions start to run high, is a bad idea. Smaller PSAPs may have a tradition of announcing changes by email or written orders. Frank, candid communication from trusted middle managers is the least traumatic method for change program deployment. To facilitate future change efforts, direct 80 percent of the organizational resources and efforts toward routinely cultivating supervisor expertise, competence, and trusted status with their subordinates. Senior management should be visible and approachable at key junctures.

The overall organizational tone should target support of continuous change and development. Information flow to employees, with a high degree of empathy for their circumstances and interests, helps to sustain the changes proposed.

Sustaining change

Sustaining organizational change has at its root an interest in preventing the typical phenomena of slippage of change programs back to the familiar and comfortable state. Buchanan et al. (2005) define sustainability as the processes in which new "improvement trajectories" continue for an appropriate period, consistent with the circumstances of the change.

Timeframes and milestones

Sirkin et al. (2005) identifies four factors influencing change success and sustainability:

- Duration,
- Integrity,
- Commitment, and
- Effort.

Change program durations of four to eight months have been found to yield the most favorable result, despite tendencies to allow change attempts to drag on for one year or more. In more complex technology projects, it may not be realistic to deploy the entire change within just a few months. In such instances, it is appropriate to devise achievable milestones, before the eighth month, to establish tangible results in the first steps of the larger program of changes.

Celebrating successes

The months following implementation of the change are especially tenuous. The project team will likely be worn out by the process and eager to rush or combine steps to get the whole thing over with. Strong personal discipline will be needed at this critical juncture. Once the first milestone is reached, it is desirable to pause for celebration and to praise those involved. It is also a good time to remind what the plan calls for next, along with the expected sacrifices and benefits.

When milestones are used in major change efforts, the achievement of the first milestone can be a time for hold outs to be cynical and declare the project hopeless. The pressures and commitments made to accomplish the previous change can be gently loosened at that juncture, but vigilance will be needed to assure gains are not lost. The true diagnostic of whether the change has taken hold is when it becomes the standard way of doing things in the organization. A delicate balance must be maintained between celebrating early successes and declaring victory.

Kotter (1995) asserts that it takes five to 10 years for major changes to soak into the organization's culture. Meanwhile, holdouts and resisters will be quick to seize any opportunity to reverse the changes and return to the previous comfort zone, sometimes under the guise of tradition. Vigilance must be maintained to guard against conscious and unconscious change program sabotage and exaggerated or inaccurate reports of new systems' failure.

Management commitment

Commitment strategy addresses the senior management and local level of support needed to raise the likelihood of successful implementation and sustained change. Clear, skillful communications with local stakeholders will temper some of the inevitable reluctance felt by those impacted by a change in established practices. Frank, empathetic presentations on the sacrifices necessary, along with the potential gains, may be enough to move the majority from a state of "strongly reluctant" to merely "reluctant-willing" (Sirkin et al. 2005). Characterizations of senior management's desires can range from reluctant to neutral to "seems to want success," with highest success probability given when the change is both desired and communicated clearly.

Team

A capable project management team, ld by enlightened and inspiring leadership, is a high predictor of program success. Use of formal project management principles should be employed to provide cogent structure and facilitate communication. The tendency to heap the change duties upon people simultaneously responsible for other aspects of the ongoing operation is inappropriate. The change effort is deserving of a high level of strategic, skilled efforts aimed at answering the complex factors outlined in this chapter. Adequate staffing, based on realistic project needs will drive the integrity of the project, bringing forward the appropriate skills and traits necessary to complete the job.

Feedback loop

Management often ostracizes dissenters as disobedient, troublemakers, and obstructionists. A more enlightened approach is to view the input from resisters as an opportunity to hear what is really concerning people and use it as a chance to react to those concerns. It is a mistake to assume general silence across the organization equals acceptance. Consider resisters' input seriously, because they are likely articulating the broad areas of concern of their tacit coworkers.

A common initial group reaction to change efforts is neither positive nor negative, but rather one of guarded, silent skepticism. Input received from resisters can break through the state of "cognitive ambivalence," typical of employees taking a wait-and-see attitude, or those intent upon waiting out the changes, looking for just the right opportunity to reverse their course back to old practices (Piderit 2000).

Those who care enough to comment may intend no dissent and disharmony, but rather desire a balanced and successful eventual outcome. People engaged in what managers label as resistance often do not intend to impede programs, but rather have sincere concerns and valuable modifications to suggest. What looks to some to be resistant behavior may be intended by the actor to be supportive, within the context of their perspectives, interests, and understanding of the situation.

Workload

Worker concerns in the face of broad changes often involve scheduling, workload, skills, confidence, and saving face from embarrassing situations. It is wise to consider the additional effort required on the part of the individual worker. Ideally, no individual's workload should increase by more than 10 percent as a result of the change. Otherwise, morale suffers and workgroup conflict is likely (McBain 2006). Optimal success is expected when the burden on the individual worker is kept below 10 percent, with diminishing returns thereafter. When people are asked to increase their workload by greater than 40 percent, the chances of sustained change are greatly reduced. The 40 percent aspect clearly suggests that new work plans requiring one person to do two jobs will not be sustainable.

Timely and accurate information is important to workers implementing new technologies, with emphasis on basic daily routine and employee interests and their needs for comfort, productivity, and safety. Managers also may appear to be stubbornly change-resistant, yet their micro-level concerns may revolve around insufficient input into designing the new systems. With careful integration and attention to interests, Sirkin et al. (2005) find that most managers would support change, even if it means more work, providing they have the necessary tools and a forum to address the changes.

Prioritized attention to *information, participation,* and *trust* are key to sustained change (Van Dam et al. 2008).

Trust and forgiveness

The cultivation of trust before, during, and after the change is deployed is a high predictor of successful, sustained organizational change. The best chances for sustained change involve the cultivation of close, supportive relationships between leaders and their subordinates. Broad cultivation of trust within the organization is an important precursor to change management, the foundations of which ideally need to be strong before the change is proposed.

One of the areas of trust establishment and maintenance worthy of considerable effort is that of reconciliation and repair of damaged relationships. People working in information technologies and PSAP settings for any period are likely to encounter situations whereby squabbles, turf-battles, and misunderstandings create residual hard feelings and impediments to trusting environments (Tomlinson et al. 2004; Ford et al. 2008). Counterintuitively, empathy and capacity to forgive are helpful in preparing workforces to sustain through the turbulence of organizational change (Yeats and Yeats 2007).

Forgiveness is not a commonly discussed behavior in the change management literature, but gestures leading to forgiveness prevent and cure disharmony, which improves general workplace atmosphere and may even negate other neurological impairments not yet identified.

While there may be a tendency to ignore such "soft" factors and to focus instead on the technological aspects of procurement and deployment, it is useful to remember that those involved will bring with them feelings of how they have been treated in the past. Such latent hard feelings can be directed to the organization in general, as well as the specific people involved in the change attempt. The project may become a lightning rod for a host of latent concerns from the past. The cultivation of a strong trusting environment is essential for a number of broad organizational needs, not just the change process. Although some changes are truly needed immediately to address a crisis that just cannot wait, it would be helpful to strategically ask before deployment if the new system can be delayed until organizational trust and relationship repair can be conducted.

Preventing slippage

Some small portion of the organization may remain entrenched in old practices and customs, sometimes waiting months or even years to return to the previous state of affairs. It is natural over time for the agenda to shift to other priorities, project team members move on to other endeavors, and those remaining on the team get distracted with new priorities. Yet, people intent upon reversal to the preferred earlier state may use subtle small steps on the way to complete return to their comfort zone.

The change debate should be transparent and forthright throughout the process. A degree of agreement about process should be achieved at the beginning of the change initiative and consensus reached on when issues will be periodically revisited. During the sustainment phase, stakeholder interests should be revisited and assessed for future programmatic modifications. It is wise to bring all parties back together, post-deployment, to explain, support, and hear concerns.

Another sustainability notion is to acknowledge the sense of loss people feel for the old ways of doing things, giving those affected a brief chance to "grieve" the sense of loss for the old ways before moving on to the new methods.

Regrets?

A final thought about change sustainment is in order. The heavy investment in any major technology project may compel CIOs to stick with the formula beyond a point of diminishing returns. It can be very difficult to filter minor issues and challenges and realize when a critical mass of hopelessness is reached.

Distraction and resources drained by a failing effort will preclude staff from acquiring new skills and competitive advantages needed for organizational growth. Stubbornly sticking with the change elements is ill-advised after an obviously botched deployment, or changing external factors make the original assumptions inappropriate or outdated. Such circumstances may be cause to revisit whether the program can or should be salvaged. Patience will overcome pressure to return to old familiar ways, but at some point, it may be prudent to have the courage to call the whole thing off. Under such circumstances, CIOs should realistically consider intentionally allowing some change efforts to decay (Buchanan et al. 2005).

Summation

The long history of change effort failure in all venues drives a need for a careful, strategic change program before jumping to make major reconfigurations. Change efforts are difficult to deploy and sustain, even under relatively ideal circumstances. Change can be disruptive and difficult, especially when loss of legacy programs and deep traditions challenge organizational identity. Public sector management policies play better to older notions of top-down change directives than more enlightened, contemporary views of teamwork, collaboration, and worker sensitivity. These constraints should not be enough to label public sector change efforts as hopeless, but rather places even more importance on strategic deployment, management support, vision creation, stakeholder integration, communications efforts, and change sustainability initiatives.

Gardner (2004) recommends the following ingredients for a change process, which serve as a good summary of change program sustainability benchmarks:

1. **Reason**—Cultivate belief in the need for change and the necessity of proceeding.

2. **Research**—Gather facts to be articulated.

3. **Resonance**—Appeal to the affective needs of those involved.

4. **Redescriptions**—Find several angles and perspectives with which to tell the story.

5. **Resources and Rewards**—Devise appropriate incentives.

6. **Real World Events**—Tie the change into existing concerns; be prepared to capitalize on major external crises.

RONALD TIMMONS is the director of Public Safety Communications in Plano, Texas. The department answers 911 calls and dispatches first responders for the city of 270,000 people, and administers a regional public safety radio system. Ron has 24 years of first responder experience in firefighting and Emergency Medical Services, including service as chief of a New York State fire department. He also has 13 concurrent years of experience as an adjunct assistant college professor. Ron has a bachelor's degree in Fire Service Administration and a master's degree in Public Administration from the State University of New York. Other graduate degrees include a master's in Homeland Security and Defense from the Naval Postgraduate School in Monterey, California, and a Ph.D. in Public Affairs from the University of Texas at Dallas.

REFERENCES

Buchanan, David, Louise Fitzgerald, Diane Ketley, Rose Gollop, Jane Louise Jones, Sharon Saint Lamont, Annette Neath, and Elaine Whitby. 2005. No Going Back: A Review of the Literature on Sustaining Organizational Change. *International Journal of Management Reviews* 7(3): 189-205.

Fernandez, Sergio, and David W. Pitts. 2007. Under What Conditions Do Public Managers Favor and Pursue Organizational Change? *The American Review of Public Administration* 37(3): 324-341.

Ford, Jeffrey, Laurie Ford, and Angelo D' Amelio. 2008. Resistance to Change: The Rest of the Story. *The Academy of Management Review* 33(2): 362-377.

Freud, Sigmund. 1984. Unconscious Feelings-Part III. *On Metapsychology: The Theory of Psychoanalysis* 11, 341-406. Pelican Freud Library, Harmondsworth.

Gardner, Howard. 2004. *Changing Minds: The Art and Science of Changing Our Own and Other People's Minds.* Boston, MA: Harvard Business School Press.

Griffith, John. 2002. Why Change Management Fails. *Journal of Change Management* 2(4): 297-304.

Higgs, Malcolm, and Deborah Rowland. 2005. All Changes Great and Small: Exploring Approaches to Changes and its Leadership. *Journal of Change Management* 5(2): 121-151.

Kettl, Donald F. 1998. *Reinventing Government: A Fifth-year Report Card.* Brookings Institution. Washington, DC.

Kiel, L. Douglas. 1989. Nonequilibrium Theory and Its Implications for Public Administration. *Administrative Review* 49(6): 544-551.

Kiel, L. Douglas. 1994. *Managing Chaos and Complexity in Government: A New Paradigm For Managing Change, Innovation and Organizational Renewal,* 173-199. San Francisco, CA: Jossey-Bass.

Kotter, John P. 1995. Leading Change: Why Transformation Efforts Fail? *Harvard Business Review* (March/April): 2-9.

Larkin, T.J., and Sandar Larkin. 1996. Reaching and Changing Frontline Employees. *Harvard Business Review* (May/June): 95-104.

Lewin, Kurt. 1947. Frontiers in Group Dynamics: Concepts, Method and Reality in Social Science; Social Equilibria and Social Change. *Human Relations* 1(5): 5-41.

Light, Paul. 1998. *Sustaining Innovation: Creating Nonprofit and Government Organizations That Innovate Naturally.* San Francisco: Jossey-Bass Publishers.

McBain, Richard. 2006. Why Do Change Efforts So Often Fail? Manager Update. *The Institute of Chartered Accountants in England and Wales* (May): 13-17.

Piderit, Sandy Kristin. 2000. Rethinking Resistance and Recognizing Ambivalence: A Multidimensional View of Attitudes Toward An Organizational Change. *Academy of Management Review* 25(4): 783-794.

Robertson, Peter, and Sonal Seneviratne. 1995. Outcomes of Planned Organizational Change in the Public Sector: A Meta-Analytic Comparison to the Private Sector. *Public Administration Review* 55(6): 547-458.

Sirkin, Harold, Perry Keenan, and Alan Jackson. 2005. The Hard Side of Change Management. *Harvard Business Review* (October): 109-118.

Thompson, James R., and Shelley L. Fulla. 2001. Effecting Change in a Reform Context: The National Performance Review and the Contingencies of "Microlevel" Reform Implementation. *Public Performance and Management Review* 25(2): 155-175.

Tomlinson, Edward C., Brian R. Dineen, and Roy J. Lewicki. 2004. The Road to Reconciliation: Antecedents of Victim Willingness to Reconcile Following a Broken Promise. *Journal of Management* 30: 165–187.

Van Dam, Karen, Shaul Oreg, and Birgit Schyns. 2008. Daily Work Contexts and Resistance to Organizational Change: The Role of Leader-member Exchange, Development Climate, and Change Process Characteristics. *Applied Psychology* 57 (2): 313-334.

Van de Ven, Andrew H., and Marshall Scott Poole. 1995. Explaining Development and Change in Organizations. *The Academy of Management Review* 20(3): 510-540.

Vince, Russ, and Michael Broussine. 1996. Paradox, Defense and Attachment: Accessing and Working with Emotions and Relations Underlying Organizational Change. *Organization Studies* 17(1): 1-21.

Yeats, Rowena, and Martyn F.Yeats. 2007. Business Change Process, Creativity and the Brain. *Annals of the New York Academy of Sciences* 1118: 109-121.

Weick, Karl E., and Robert E. Quinn. 1999. Organizational Change and Development. *Annual Review of Psychology* 50: 361-386.

Chapter 20: The Power of Video to Save Lives

BY BOB STANBERRY AND JENNIFER BREMER

Today's global economy and the use of the Internet has created new threats, such as cyber attacks, cyber bullying, and increased racial and religious tensions that require immediate action. However, these new threats have not replaced the old ones, such as natural disasters, catastrophic accidents, property crime, murders, and assaults. These new threats simply have been added to an already depleted and overworked emergency and public safety infrastructure. As a result, there is an even greater need for real-time communication among government agencies, police forces, fire departments, EMS, and other personnel who help ensure the safety of residents. A delay in communication caused by disparate systems that cannot "talk to each other," as well as the lack of true situational awareness caused by the inability of most systems to relay real-time video and text updates, can cost lives.

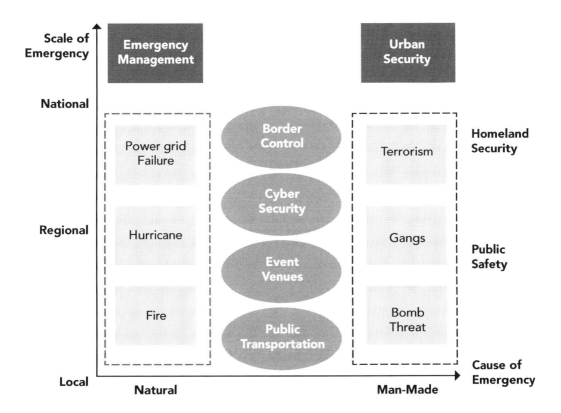

Figure 1. The varied scale and security challenges of public safety emergencies.

In a world of new and expanding threats, coupled with declining budgets, public safety agencies are being forced to evaluate the ways in which they serve and protect, often having to pick and choose between vital programs and services. For agencies, the result has been reduction in forces and the delay of new programs. But, it is not all doom and gloom. There is a way to work within this budgetary climate (that is not going to get any better any time soon, by the way), but it requires a fundamental shift in the way in which public safety agencies view and use technology. By integrating networking and communications technologies across the organization and taking advantage of telepresence, integrated voice communications over IP, and mobile collaboration technologies, agencies can ensure uninterrupted delivery of services and communication, keep more police on the street, and lower overall operational costs.

Providing face-to-face collaboration over video allows public safety officers and officials to exchange ideas for improving public safety practices, while cutting the costs of travel and training. It allows for immediate situational awareness and e-warrants that speed response times and streamline the evidence-gathering process. It even can protect law enforcement officers by allowing for remote arraignment instead of expensive and risk-prone transportation of prisoners. Additional services enabled by mobile video capabilities, such as remote interpretation, can aid officers in obtaining vital witness testimony and also assisting in the arrest of suspects who do not speak the local language.

Technology-savvy agencies are finding new and better ways to incorporate telepresence and video capabilities into everyday public safety activities, such as:

- **E-warrants**—Remote access to courts from precinct or squad car provides faster response and resolution. It improves the entire process of issuing, servicing, and clearing warrants by providing immediate access to resources.

- **Arraignment and appearance**—Detainees appear via interactive video directly from their holding facility for motions and testimony, increasing responsiveness and speed of justice while reducing security risks and costs of transfer. Law enforcement agencies reduce confrontation and escape possibilities as well as administrative costs, and can connect detainees to multiple court facilities.

- **Language interpretation**—Immediate access to interpreter services facilitates faster response to medical, legal, and detainee management. Multiple jails can use a pool of interpreters regardless of location. Costs for interpreters are reduced, and services are improved with live, face-to-face interaction between inmate and interpreter. On-scene services are accessed from the squad car by the responding officer for immediate interviews and faster resolution.

- **Situational awareness**—Video and instant field status can be delivered to and from the scene regardless of geographic location. First responders receive and send information instantly with more effective alerting, information exchange, and visual access to situations for improved crisis response.

- **Administrative meetings**—Law enforcement officials, staff, and administrators connect and collaborate together and with officials regardless of location. Training and planning meetings take place without the need for travel, reducing costs and time required with travel and logistics.

- **Education**—Law enforcement officers and staff can access educational services and programs instantly and at reduced costs to the agency. Personnel can stay inside their districts to maintain response times while allowing access to continuing education and certification programs.

In addition to enhancing everyday public safety activities, technology also can protect vital information, establish contingencies that allow staff to perform functions under extreme conditions, and maintain access to critical resources among agencies and residents through interoperable communications.

A disaster can result in suspended services and loss of access to data, applications, and work facilities for government agencies. Every disaster brings a wave of chaos; only those agencies that are best prepared can respond quickly and effectively to bring disparate elements together and reestablish order to keep critical functions online. Government agencies with disaster recovery plans and necessary technology infrastructures in place will fare better with greater continuity of services.

Government agencies must tie all the components needed for public safety systems together. A "network-as-a-platform" approach—along with open, standards-based architectures—enables agencies to:

- **Cut costs**—The existing infrastructure can be reused and integrated so that you can use your budget to enhance, not replace. A network-based approach also means that security systems can be used for much more than just security, ensuring better ROI.

- **Make intelligent decisions**—Use of collaboration technologies delivers information—including voice, video, and data—in real time to the relevant personnel, wherever they may be. This paradigm allows personnel to get and share the information they need for a faster and fully collaborative response.

- **Scale**—A unified, IP-based security system provides the foundation for integrating with other systems and devices. It allows best-in-class security and communications components from multiple vendors to be combined. This combination, in turn, helps ensure scalability and can greatly enhance the situational awareness of personnel.

- **Reduce training needs**—Integration with existing installed systems and the ability to work with your preferred tech partners helps to ensure that less training is needed for your staff to get acquainted with new systems.

To confront contemporary threats and risks, successful strategies focus on collaboration. Integrating all the solution components is essential. People, processes, training, and planning are all part of effective disaster preparedness. Mobile collaboration, including videoconferencing, is at the center of all of this preparedness—connecting people anytime, anywhere, on any device.

Videoconferencing solutions, including telepresence, place people at the center of the collaboration and video experience, empowering them to work together in new ways to transform business, accelerate innovation, and do more with less. The Cisco TelePresence® Solution represents the next generation of videoconferencing, where everyone, everywhere can be face-to-face and more effective through the most natural and lifelike communications experience available.

All of this information sounds great, but does it work in the real world? Following are a few examples of how government agencies are using communications solutions to help ensure public safety around the country:

Missouri State Highway Patrol

The Missouri State Highway Patrol (MSHP) is an internationally accredited law enforcement organization with 1,200 sworn officers, including officers from the Gaming and Drug and Crime Control divisions, and more than 1,200 uniformed civilians and support staff. MSHP is responsible for enforcing traffic laws on Missouri's 33,000 miles of state-maintained highways, as well as motor vehicle inspections, commercial vehicle enforcement, driver's license examination, criminal investigations, criminal laboratory analysis and research, public education, gaming enforcement, law enforcement training, and more. MSHP works in conjunction with state, county, local, and federal agencies in the coordination of emergency and nonemergency communications.

"Ensuring interoperability and data exchange between law enforcement agencies within a state, especially during a crisis situation, is the biggest challenge for law enforcement agencies today," says Captain Kim Hull, director of the MSHP communications division. "Interoperability is critical for public safety success." MSHP identified several statewide events and natural disasters where communication had been a problem. Because of disparate communication technologies, especially over the radio between state and local agencies, MSHP recognized a need for change.

Figure 2.

Missouri deployed its state-of-the-art Network Emergency Response Vehicle (NERV). This vehicle is a mobile communication center, supported by a highly trained team that is designed to establish interoperable communications in emergency situations. With the NERV, the State of Missouri did not need additional vendors to establish statewide communication during the tornadoes in spring 2011. In this example, the NERV not only saved the state money and time because state officials were ready for the tornadoes, but it also protected the lives of locals living across Missouri by enabling communication with emergency responders and residents.

State of Indiana, Department of Homeland Security

When disaster strikes, the Emergency Operations Center (EOC) becomes the Indiana Department of Homeland Security's (IDHS) support base for all state multiagency deployment teams. Because of these natural disasters, IDHS determined the need for a reliable, real-time communications system that would be independent of ground infrastructure.

"The ability to deliver real-time video and pictures from the heat of the battle to the EOC gives us a better grasp of the scenario and allows for more efficient deployment of resources," says Shane Booker, the response director for the IDHS.

To ensure connectivity throughout the state, four satellite-based vehicles were constructed to support field operations and provide information to and from the EOC: two mobile satellite office solutions (SOS) units, a mobile command center (MCC), and an incident response vehicle (IRV) were developed. The entire fleet was designed such that each vehicle can serve as an alternate for the state EOC or a county EOC.

Each deployment has showcased functions that, while available inside the EOC, were previously nonexistent on scene, including the WebEOC, dial tone, and live video sent back to the EOC. Continued use of these vehicles to support events and training scenarios will help ensure IDHS and IOT will be ready to provide rapid response when the next emergency or disaster occurs in Indiana.

Bexar County, Texas

Bexar County envisioned a safer way of accomplishing its legal and judicial processes. For example, in the Central Magistrate, some judges did not have a holdover cell in their courtrooms, increasing the risk associated with physically bringing in inmates for legal proceedings. In some courtrooms, holding inmates is particularly risky, given the proximity of the courtroom to an exit, the crowded nature of the room, and the general hostility of the people in the courtroom. Bexar County, and all of its courts, needed a system that would enable secure, unified communications and collaboration between justice and law enforcement departments, as well as partner organizations.

In total, Cisco established 20 Cisco TelePresence endpoints throughout Bexar County to streamline communications for court processes and facilitate collaboration across different justice and law departments. Five of the video teleconferencing (VTC) units were intentionally installed in the county jail, and all are in secure "inmate-proof" encasings to facilitate legal communications between lawyers and their clients (for example, for pretrial hearings).

Cisco TelePresence videoconferencing also has succeeded in reducing risks to public safety in the county, primarily by eliminating the unnecessary transportation of dangerous and/or mentally unstable inmates between courtrooms and the state hospital. Specifically, judges use videoconferencing for mental-competency hearings. Similarly, by allowing for virtual communications and subsequently decreasing the use and need for holdover cells, videoconferencing has successfully reduced courtroom incidents. Court proceedings can unfold with lower risks, including flight risk, for all parties. Another poignant result has been that, by facilitating remote testimony, videoconferencing has decreased the need for confrontation in court proceedings, particularly in the Bexar County Children's Courts during child abuse and neglect cases.

Although each of the different courts in Bexar County uses a Cisco TelePresence system in different capacities, the end results are the same for all: increased efficiencies, increased public safety, and saved resources. Cisco videoconferencing technology has, so far, saved Bexar County hundreds of thousands of dollars stemming from the resulting efficiencies. Over the next few years, the videoconferencing system is expected to pay for itself because of the significant savings (for example, by eliminating expert testimony fees, which include transportation, board, and room).

Telepsychiatry

Most county jails across the United States cannot legally process through the jail system arrested individuals who claim a psychiatric emergency without first receiving psychiatric and/or medical clearance. This process was established to protect the safety and well-being of multiple parties, including officers, other inmates, and mentally ill individuals. Individuals arrested during "off hours," or at night, could not receive medical evaluations until psychiatric personnel were available the next morning. In extreme cases, prisoners arrested on Friday evenings would be admitted to the hospital and continuously guarded by police officers for the entire weekend while waiting for a psychiatric evaluation the following Monday. The processing regulations, albeit necessary, negatively affected multiple parties. Much-needed police labor was lost as officers were required to watch prisoners in the hospital, care for individuals truly requiring medical attention was delayed, and falsified claims backed up the system—all while consuming already limited police budgets.

Deploying Cisco TelePresence systems in jails and providing 24-hour access to psychiatrists allows police departments and correctional institutions to offer psychiatric evaluations to arrested individuals immediately. This paradigm eliminates the need for an officer to transport the suspect to the hospital and wait with him/her during the psychiatric evaluation. Conducting psychiatric evaluations with Cisco TelePresence conferencing complies with all of the standards for healthcare and establishes a precedent for other jails and/

or rural hospitals that also could use videoconferencing technology to respond to the increasing need for immediate access to care.

Telepsychiatry increases efficiency in regard to processing inmates, because it now requires less time and local manpower to facilitate claims. Before inmate processing was lengthy and tedious, taking a total of six to eight hours (not counting weekend or overnight incidents). By contrast, with the new system in place, the process now takes less than an hour.

A stay for an inmate at a hospital over the weekend costs approximately $8,000 to cover room and board and 24/7 care, not including overtime pay for police officers required to watch these inmates. It also reduces the threat to the public by ensuring that the arrested individual does not have a chance to escape or harm others in a public place. By eliminating hospital psychiatric evaluations, state and local governments saved taxpayers a significant amount of money.

City of Hayward, California

The Hayward Fault is a geologic fault zone about 37 miles long that lies mainly along the western base of the hills east of the San Francisco Bay. Parallel to its more famous (and much longer) neighbor to the west, the San Andreas Fault, the Hayward Fault can generate significantly destructive earthquakes right through the center of Hayward. "The Fault is on a 150-year cycle of large earthquakes, with the last one occurring approximately 152 years ago," says Clancy Priest, chief information officer for the City of Hayward. "With

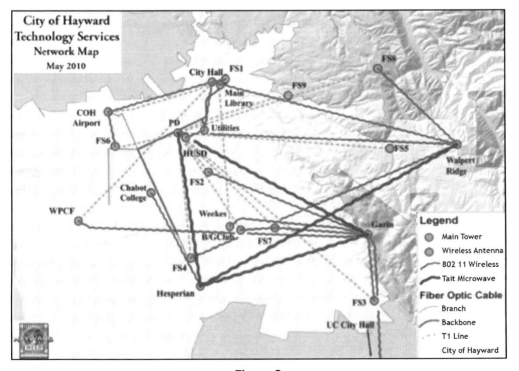

Figure 3.

this delay in the Fault's pressure release, we are well overdue for a major earthquake in our city. We had to look at the business continuity of our city in the event of a major earthquake. How would the city stay connected? How would we function? When I became CIO in 2002, I became dedicated to answering these questions."

The city looked at its business continuity plans in the event of an incident and determined how the city would stay connected. It concluded the most effective way to connect city sites after a natural disaster was via an 802.11 wireless bridge. With a pervasive wireless network, the city's first responders, police, hospitals, and other essential resources can stay connected and collaborate to save lives and property during an earthquake of any magnitude. For example, city officials will now have direct access to emergency responders and federal officials during an emergency to ensure that all persons agree on a plan of action.

The city is also looking at Cisco TelePresence systems on the network as well in the fire department, emergency operations center, and nine fire stations. "Our goal is to use the advanced videoconferencing service for briefings, trainings, and emergency updates," says Priest.

As these examples clearly demonstrate, technology is the key for helping state and local governments handle any challenge—from natural disasters to psychiatric evaluations—all while staying within the parameters of a tight budget and even cutting operating expenses. At its core, today's collaboration technology helps save lives and money by allowing for officials to stay in communication no matter what.

BOB STANBERRY, who leads Cisco's Connected Justice Solution Business Development team, has more than 15 years of experience in law enforcement, justice, safety, and security in both the public and private sectors. He has experience in state and local law enforcement, and served as a police grant liaison and chief of police. Bob has served as a consultant for the Justice Department, OJJDP and has worked with agencies across the United States, helping them build their capabilities with a focus on technology. He also has worked in the private sector where he was focused on solutions that monitor critical infrastructure, schools, and private organizations.

JENNIFER BREMER has more than 15 years of experience in the public sector market and specializes in the use of technology in public safety. A graduate of North Carolina State University and East Carolina University, Jennifer has focused on leveraging technology to communicate and educate public sector leaders on the social and budgetary benefits of IP and video-based solutions and architectures. Jennifer is actively engaged in key associations; she is a member of the International Association of Chiefs of Police Community Policing Committee, the Digital Communities Law Enforcement Task Force, and the NACo Green Government Committee. Currently, Jennifer is the Director of Marketing for Cisco's State and Local Government and Education Markets.

Chapter 21: Integrated Risk Management— A Public Safety Perspective

BY MIKE KENEL

Introduction

State and local public safety agencies often are assigned responsibility for managing homeland security and emergency management initiatives within their respective jurisdictions. Funding and guidance for the initiatives come primarily from the Department of Homeland Security (DHS) in response to presidential directives or enacted legislation. A central tenet regarding DHS funding and authorization of state and local expenditures is that such investments should be made following a risk-based assessment of the threats and hazards within the jurisdiction.

Despite DHS attempts to provide state and local communities with risk-based scoring methodologies (e.g., CARVER, RAMCAP, MMSHARP), no one method has gained widespread acceptance. Overly simplified methods leave too much to individual interpretation, and planning teams become aware that final scoring results are based more on conjecture rather than fact. In contrast, overly complex methodologies require a level of risk management expertise typically not found on local planning teams. The shortcut that many jurisdictions have had to take has been to simply assign some level of "importance" based on gut instincts or best professional judgment. What remains a priority need is development of a risk assessment methodology that ranks the relative risks associated with the need to prevent, protect, respond, recover, or mitigate the negative consequences associated with the greatest risks facing local jurisdictions.

Benefits

There are a number of reasons why state or local public safety officials might want to recommend risk management principles be integrated into their respective organizational structures, including:

- The ability to identify and address any new risks well in advance of the community enduring the negative consequences of such risks,

- The determination of appropriate risk reduction options based on an alternatives analysis that looks largely at the costs and benefits of each option,

- The generation of meaningful data in support of budget requests to the community's leadership, as well as grant requests from state and federal governments, and

- Documented evidence that public officials demonstrated "due diligence" when asked to address questions from the public or news media on the community's level of preparedness.

The purpose of this article is to provide state or local public safety officials with guidance on establishing an integrated risk management system across all levels of government and to provide guidance on the design of a risk assessment process that safety officials can use periodically to address state or local community risks.

Key definitions and description

In 2008, DHS published guidance (DHS Risk Lexicon, September 2008) promoting the use of consistent terminology to improve communications and understanding regarding risk-related issues. Figure 1 provides a number of key definitions that help define some of the concepts presented in this article. These particular definitions also support the following simplistic description of how an integrated risk management system is implemented within an organization.

Event or Incident
A natural disaster, act of sabotage, or technological failure that may cause harm.

Risk
The potential for an unwanted outcome resulting from an incident, event, or occurrence, as determined by its likelihood and the associated consequences.

Key Assets
A person, structure, facility, information, or material that has significant value to the community.

Essential Functions or Services
A process, capability, or operation performed by an asset, system, network, or organization that is critical to the community.

Risk Governance
The personnel, practices, processes, and mechanisms concerned with how risks are to be analyzed, managed, and communicated.

Hazard Assessment
The identification of natural or man-made sources of harm or difficulty.

Threat Assessment
The process for identifying or evaluating entities, actions, or occurrences, whether natural or man-made that have the potential to harm life, information, operations, the environment and/or property.

Vulnerability Assessment
The process for identifying physical features or operational attributes that renders an entity, asset, system, network, or geographical area susceptible to hazards.

Risk Assessment
The collecting of information and the assigning of values to perceived threats and hazards in order to develop priorities, compare courses of action, or inform decision makers.

Figure 1. Key definitions.

Integrating a risk management system within a local jurisdiction starts by identifying the critical infrastructure, key resources, and essential services that are needed by the community to sustain its vitality and opportunities for growth. Next, an analysis is completed of the threats and hazards potentially directed against the community's infrastructure, resources, and services, including the identification of any vulnerability to the threats and hazards. Public safety officials now could begin a formal assessment of the community's risks, calculating the risk and ranking the risks to a defined risk scoring methodology. Starting with the highest ranked risks, safety officials would conduct an alternatives analysis, which would allow community leaders the opportunity to decide either to accept the risk (basically do nothing), transfer the risk (as with insurance), avoid the risk (agree not to implement the action creating the risk), or lessen the risk (largely through the adoption of a capability or countermeasure that provides the community with the means to prevent, protect, respond, recover, or mitigate the risk). The authority, responsibility, processes, and means for conducting these activities is defined in a set of documented policies and procedures that form the basis of a risk management system.

Risk governance

Risk governance refers to the personnel, practices, processes, and mechanisms within an organization that will be involved with assessing, analyzing, and managing the organization's risks. One of the first steps in implementing a risk management system is for management to issue a risk policy designating a risk manager whose role is to develop and implement the system. Also, the risk policy often will designate a risk management planning team composed of representatives from various agencies and departments within local government. The planning team's role is to support the risk manager in implementing the risk management system and to assist the risk management in operating the system once it is developed.

DHS published the results of a survey on risk management practices (Risk Management Practices in the Public and Private Sector, Executive Summary, September 2010) within the private and public sectors, bringing to light some of the challenges in integrating risk management practices within an organization, including:

- The need for the organization's leadership to not only endorse the risk management program but consistently practice risk-based decision making in order for the program to be successful.

- The need to move beyond mere periodic risk assessments, to where risks are tracked on a more regular basis, providing an early warning of any changes in risk exposure.

- The need for risk metrics, such as those that might be developed as part of a risk scoring methodology, to be kept relatively simple and understandable so that decision-makers can effectively manage those risks and communicate information regarding those risks both within and outside the organization.

Thus, the DHS concept for governing risk is one where state or local senior leadership endorses the adoption of risk management principles across all levels of the organization and uses defined metrics in analyzing and communicating those risks as needed. In contrast, actual risk governance within many local communities tends to be less well-defined. Consider, for example, the budgeting process and how it relates to addressing a community's risks to natural disasters, acts of sabotage, and technological failures. In many communities, budgets are allocated across departments where each department addresses risks in a segregated manner rather than in a more holistic or enterprise approach where a single prioritized list of the community's risks is produced. Often what drives a department's risk management budget is not an analysis of the risks, rather what historically has been budgeted (give or take some percentage based on whether the community's revenue is increasing or decreasing in a particular year). How the department then allocates the provided funds is based more on "gut instinct" as to what is the most appropriate allocation of funds rather than a systematic evaluation of the risks where the decision makers learn more about the pros and cons of their decision.

Fundamentals

In early 2011, DHS published its guidance on how risk assessments are to be performed and the means for institutionalizing risk management practices within an organization (Risk Management Fundamentals, April 2011). Key points to consider include:

- Homeland security organizations need to not only manage external risks (natural disasters, malicious acts, infrastructure failures) but internal risks (workforce management, project costs, legal liabilities, etc.).

- Risk management embraces the prioritization of competing demands, providing decision-makers with options or choices in reducing risks to the community, as well as measuring the community's progress toward risk reduction.

- Risk management policies and procedures are not part of a standalone program, rather policies and procedures are integrated into the community's planning processes, including homeland security strategic planning, capability planning, resource planning, operations planning, and training and exercise planning.

Lessons learned

In addition to DHS guidance, the author, in implementing risk reduction programs for state and local governments, has experienced a number of practices that can enhance the effectiveness and acceptance of a risk management program.

First, consider that a number of federal programs (i.e., Hazard Mitigation Planning, Continuity of Operation and Government Planning, Critical Infrastructure Protection Planning, etc.) require that local governments assess the risks within their communities. Local communities may want to consider these programs in designing a single comprehensive risk assessment that addresses the risk identification and analysis step in each of these programs. Although the resulting risk assessment will need to be periodically updated, combining this effort will not only save considerable time and expense, but also provide an appreciated comprehensive viewing and prioritization of the risks across the community.

Second, individual members of the planning team often can bring some extreme biases regarding the outcome of any risk assessment. In some instances, individuals may attempt to influence the scoring in wanting the scores to be higher for the risks managed by their department, thinking that higher ranked risks would likely receive more funding. Others may want to see the risks their department manages scored lower, thinking they would escape additional work or perhaps be accused of not properly addressing the risks in the past. To overcome these biases and promote acceptance of the final scoring and prioritization, it is essential that time be taken to build consensus among the team members as to the appropriate scoring methodology. Once consensus is reached on the scoring methodology, it becomes more difficult to question the assessment's results because altering the final ranking for any one political purpose undermines the scientific integrity of the entire assessment.

Third, although DHS has proposed a number of risk-scoring methodologies, no one method has been universally accepted. Also, despite DHS promoting risk-based decision making in many of the federal programs, seldom do those programs detail the methodology to be used for scoring the risk. Risk can be calculated by qualitative or quantitative means. Quantitative risk assessments typically require considerable data collection, expertise to analyze and interpret the data, and the subsequent cost. For this reason, they are seldom used in homeland security and emergency management initiatives. In qualitative risk assessments, designations such as "low," "medium," or "high" are used to designate different levels of probability, consequence, and risk. The drawback to qualitative risk assessments is that the categorization of risk is often so simple that it lacks sufficient detail to distinguish two risk scenarios given the same designation.

The use of a semi-quantitative risk assessment is an attempt at compromising between the labor-intensive quantitative risk assessment and the overly simplified qualitative risk assessment. Semi-quantitative risk assessments are the most common type of risk assessment used in homeland security programs. Semi-quantitative risk assessments replace the "low, medium, high" designations for probability of occurrence and consequences with numeric values, typically 1 to 5 or 1 to 10. This allows for a greater number of categories to be defined, and the numeric values for probability and consequence now can be used in a mathematical equation to calculate a relative risk number more suitable for ranking.

Finally, if one considers the number of potential threats and hazards that could be directed against the public, key assets, or essential services, it becomes apparent that one could spend considerable time collecting and analyzing various risk scenarios. (A risk scenario is a hypothetical situation comprised of a threat or hazard, an impacted entity and some expression of the consequence. See Figure 2 for examples). As part of any risk assessment, communities will want to evaluate the top risk scenarios confronting the community. "Criticality" and "Reasonable Worst-Case" are two concepts that can help focus the ef-

> Floods in the downtown area result in significant property damage and a loss of revenue for local businesses.

> A pandemic outbreak leads to a 25% incidence of prolonged illness in the general population.

> A severe winter ice storm results in a regional power outage, with major areas being without power for up to 30 days.

> Cyber intrusion into the government IT financial systems cause a disruption in public subsistence payments.

> Three railcars containing a deadly gas derail and release their contents downwind over a highly populated area.

Figure 2. Example risk scenarios.

fort, saving considerable time and effort. "Criticality" refers to arbitrarily selecting a risk score above which the associated scenario is deemed important in regard to conducting an alternatives analysis on which course of action best addresses the risk. Risk scenarios are to be scored only if it is anticipated that the resulting risk score will be near or above the criticality criteria. This will avoid time spent collecting and analyzing data on risks that are not likely to be critical.

The "Reasonable Worst-Case" concept suggests that when developing the hypothetical risk scenarios, the planning team not develop a number of scenarios that differ only by degree or severity, nor would they develop risk scenarios that are so extreme as to be unlikely to ever occur. Instead, scenarios are developed that are somewhat worse-case but fall within the range of a reasonable possibility. The assumption is then made that in addressing the "Reasonable Worse-Case" scenario, the planning team is likely to be addressing any related, but less severe, scenarios.

Integrated risk management

Institutionalization of risk management principles are best achieved through the implementation of a risk management system having a risk policy, assigned responsibilities, risk-related goals and objectives, documented procedures, and performance metrics. Staff are to receive training on the procedures, and top management within public safety would review the system's performance metrics at least annually, taking corrective actions as necessary. The program's overall objective is to continuously improve on the performance metrics, constantly striving to reduce the community's risks to zero, while realizing zero risk will never actually be achieved. The resulting risk management system has been coined and promoted by DHS as an Integrated Risk Management (IRM) System. The local community staff's adherence to the documented procedures is intended to produce accuracy and consistency in the various work products produced by the system. Work products produced by the system can include risk assessments, a risk inventory, a risk map, a relative risk ranking, an alternatives analysis, program status reports, and work plans implementing the various control measures intended to actually maintain or reduce the level of risk.

With an IRM system in place, the risk manager and the risk management planning team are well positioned to complete a community-wide risk assessment.

Risk assessment

A starting point for the risk assessment can be in the identification of the community's critical infrastructure, key resources, and essential services that the community requires in order to sustain its economic growth and quality of life. Figure 3 identifies the types of assets (infrastructure, resources, and services) that the community might consider as essential.

INFRASTRUCTURES	RESOURCES	SERVICES
• Dams and Reservoirs	• Gasoline	• Law Enforcement & Fire Services
• Power Grid and Power Plants	• Propane	• Emergency Medical Services
• Water and Wastewater Systems	• Heating Oil	• Public Subsistence Programs
• IT/Computer Systems	• Electricity	• Postal and Shipping Services
• Schools and Universities	• Natural Gas	• Payroll and Taxation Services
• Bridges and Tunnels	• Diesel Fuel	• Court and Prison Services
• Hospital and Medical Facilities	• Aviation Fuel	• Emergency Notification Systems
• Air and Sea Ports	• Food	• Emergency Dispatch Services
• Manufacturing and Retail Centers	• Water	• Emergency Management Services
• Rail lines		• Public Works Services
• Stadiums and Arenas		• Mass Transit Services
• Chemical Plants and Pipelines		• IT Services
• Oil Refineries and Tank Terminals		
• Petroleum and Natural Gas Pipelines		
• Centers of Government		
• Banking and Finance Systems		
• Telecommunication Systems		

Figure 3. Potential key local infrastructures, resources, and services.

Figure 4 (see next page) identifies the types of threats and hazards that might be directed against a local community and its assets. A Threat and Hazard Assessment determines which threats and hazards are relevant to the community, and from this assessment, a number of risk scenarios can be developed that represent the greatest risks that might be directed against the community (revisit Figure 2 for examples). A listing of these risks and associated information can form the basis of a Risk Inventory — a communication and tracking device that provides over time a record of actions related to each risk. A Risk Map superimposes hazardous areas over the location of community assets and demographics to better visualize the geographic relationship between impending threats or hazards and those individuals, institutions, or structures that might be at risk.

> **Natural Disasters**
> • Earthquakes • High Winds, Tornadoes, Hurricanes • Flooding • Snow and Ice Storms
> • Prolonged Temperature Extremes
>
> **Human-induced Disasters**
> • Cyber Intrusion • Acts of Sabotage, Terrorism • Chemical or Biological Release
> • Pandemic • Explosions, "Dirty" Bombs
>
> **Technology-based Disasters**
> • Industrial Explosion • Chemical Release • Infrastructure Failure

Figure 4. Hazard considerations.

Each risk scenario will now be scored based on its relative risk. As mentioned previously, semi-quantitative risk assessments are often used in homeland security programs. A scale of 1 to 5 or 1 to 10 to represent graduated degrees of severity, intensity, or frequency is used to score the probability (P) and consequences (C) of an event. Risk (R) can then be calculated by the equation:

$$R = P \times C.$$

P also could be measured by a sum of values representative of P, such as visibility, accessibility, and vulnerability, where each could be scored on a scale of 1 to 5 or 1 to 10 and then summed as follows:

$$P_{total} = P_{visibility} + P_{accessibility} + P_{vulnerability}$$

Similarly, C (the total consequences of an event) can be calculated based on a number of factors, including impacts on human health, the environment, personal property, and the economy, such that:

$$C_{total} = C_{health} + C_{environment} + C_{property\ damage} + C_{economy}$$

Again, each of the above components in the equation can be scored on the scale of 1 to 5 or 1 to 10 and then summed to give a numerical value for the total consequences of an event (see Figure 5).

Finally, in designing a risk scoring methodology, the planning team may want to introduce for each component of probability (P) and consequence (C) a weighted value (W) that reflects their relative importance. For example, impacts on human health often are considered more important than the impacts on personal property and, therefore, are assigned a higher weighted value. The use of weighted values is important in building consensus among planning team members, as it can help resolve differences of opinion on what factors should be assigned more importance when determining risk.

Our risk equation now can be expressed as follows:

$R = P \times C$, or

$R = P_{total} \times C_{total}$, where

$P_{total} = [(W_{visibility} \times P_{visibility}) + (W_{accessibility} \times P_{accessibility}) + (W_{vulnerability} \times P_{vulnerability})]$,

and

$C_{total} = [(C_{health} \times W_{health}) + (C_{environment} \times W_{environment}) +$

$(C_{property\,damage} \times W_{property\,damage}) + (C_{economy} \times W_{economy})]$

Variations in risk scoring

There are a number of considerations the risk management planners should keep in mind as the risk scoring methodology is finalized.

First, in regards to the scoring of consequences, consideration can be given to including other types of consequences in the risk equation, including public outrage, loss of faith in the government, or a diminished quality of life.

Raw Score	Scoring Criteria	Weighted Score
\multicolumn{3}{c}{Threat Component Target Attractiveness (Weighted Value = 10)}
0	Site is of no real value to lone wolf, or domestic or international terrorist organizations	0
1	Site's loss would cause only temporary inconvenience	10
2	Site is of limited value, involving neither the loss of critical infrastructure or mass casualties	20
3	Site is a critical infrastructure site within the jurisdiction or provides an essential function or service	30
4	Site is of national or regional importance and, therefore, of relatively high value	40
5	Site's loss would create public outrage or compromise quality of life on a long-term basis	50
	Other candidate criteria include: Adversary Capabilities, Adversary Proximity, Historical Frequency (Natural Disasters)	

Figure 5. Example risk scoring table.
(continued on next page)

CHAPTER 21: INTEGRATED RISK MANAGEMENT

Raw Score	Scoring Criteria	Weighted Score
Vulnerability Component Site Visibility (Weighted Value = 5)		
0	Site is underground or concealed, and site's purpose or mission is largely unknown	0
1	Site is in a relatively remote location typically known only by local residents	5
2	Site is typically acknowledged on highway maps, and the public is generally aware of its existence	10
3	Generally known as an important site, with directions to site being provided on nearby highways	15
4	Site is highly visible and a well-known entity whose location is often promoted to the public	20
5	Site's location is well-known and nationally or internationally recognized for its importance	25
	Other candidate criteria include: Warning Duration, Infrastructure Design, Population Density, Natural and Social System Resilience	
Consequence Component On-site Property Damage (Weighted Value = 10)		
0	Infrastructure Repair/Replacement costs less than $100,000	0
1	Infrastructure Repair/Replacement costs over $100,000 but less than $1,000,000	10
2	Infrastructure Repair/Replacement costs over $1 million but less than $10 million	20
3	Infrastructure Repair/Replacement costs over $10 million but less than $100 million	30
4	Infrastructure Repair/Replacement costs over $100 million but less than $1 billion	40
5	Infrastructure Repair/Replacement over $1 billion or an irreplaceable Historical Site or Landmark	50
	Other candidate criteria include: Human Health Impact (Site and/or Public), Environmental Impact, Economic Impact, Collateral Property Damage, Revenue Loss, Business Disruption, Job Loss	
Security and Response Component Site Physical Security (Weighted Value = 10)		
0	Gates open, no guards or cameras, no intrusion detectors	0
-1	Security fence and controlled access to area	-10
-2	Gates, access control, video surveillance to non-manned or inadequately manned station	-20
-3	Identified vulnerabilities and have concentric circles of different securities about vulnerable locations	-30
-4	Able to detect intruder and summon local law enforcement before negative consequences inevitable	-40
-5	Able to detect intruder and summon adequate deterrent force before negative consequences inevitable	-50
	Other candidate criteria include: Cyber Security, Continuity of Operations, Continuity of Government, Response Capability, Recovery Capability, Training, Drill and Exercises, Targeted Capability Gap Closure, Interdependencies	

Second, in order for a risk assessment to be performed accurately, it is important to clearly pre-define the basis of each numerical scale. The more detail that can be provided in the risk assessment procedure regarding what each numeric value represents, the better. If too much is left to interpretation, disagreements can arise among planning team members as to how something is actually scored. For example, when scoring the economic impact of an event, without sufficient detail, questions can arise on exactly how economic impact is to be scored, including:

- When evaluating economic impact, does the planning team look at local, regional, or state impacts, or all three?

- What types of revenue loss are to be included (e.g., taxes, salaries) in determining economic impact?

- Does the team look at just the lost revenue from the businesses that were physically damaged, or do they go back into the supply chain and look at how suppliers were impacted in terms of their lost revenue?

Again, having the planning team answer those questions during the design of the risk scoring methodology will help build consensus and ownership into the final methodology, making acceptance of the final risk ranking more likely.

Third, keep in mind the types of questions the risk assessment is intended to answer. The risk assessment design may be influenced if the types of risks that are to be evaluated go beyond natural disasters, technological failures, and acts of sabotage. Although public safety officials may be comfortable limiting the risk assessment to disaster-related risks, key decision makers (e.g., mayors, city managers, commissioners) that will use the risk assessment results may want to expand the risk management system to include financial, safety, legal, and other types of risks so as not to exclude them from the community's overall risk management picture.

Fourth, it has been proposed by some that the probability score be eliminated from the risk calculation for a number of reasons. With regards to terrorist-based risk scenarios, given that the probability of any one community being attacked is very low (i.e., 0.000001 or less), scoring on the basis of a 1 to 5 or a 1 to 10 value might seem unjustifiable, as they would over estimate the probability. Also, terrorist threats appear to be constantly changing. At one time or another, water systems, malls, the power grid, nuclear plants, and shipping ports all became the "threat of the week." For a local community that had just ranked its risks and began allocating resources to address those risks, the constantly changing threat picture can be frustrating. (One option is to reserve some resources to address new or unexpected risks and to have in place a documented process for staff to evaluate these new threats on their own merit and then to present the findings to the community's decision makers unencumbered by previous assessments.)

Finally, a community may want to consider adding a third component to the risk scoring methodology, one where recognition is given to already existing countermeasures.

Consider two identical power plants serving very similar communities. Suppose one facility has made considerable investments in physical security, staff training, and exercises directed at response and recovery, while the second facility has made very few investments. Typically, you would expect the unprotected plant to be higher in the risk ranking. This can be accomplished by reducing the risk score of the protected plant by having a security component (again rated on a 1 to 5 or 1 to 10 scale) comprised of negative values. Thus, the greater the amount of security that has been implemented, the lower the risk score.

Alternatives analysis

Once the relative risk scores for each risk scenario are determined, the scenarios are ranked. The highest-ranked risks are acted upon first in order to allocate limited resources in a manner that results in the greatest overall reduction in risk for the dollar expended. An alternatives analysis is performed in order to provide the local community decision makers with the best information possible upon which to make a decision. An alternatives analysis looks at four basic manners by which the community may decide to address a risk:

1. Do nothing and accept the risk particularly if the level of risk is tolerable.

2. Avoid the risk by removing the asset or service upon which the risk is based.

3. Transfer the risk either through insurance or by agreeing with an outside partner to share the risk.

4. Control the risk by implementing countermeasures that can achieve some level of risk reduction.

Controlling risks are often achieved by adopting one or more capabilities that are directed either at preventing the event from occurring, directed at protecting a key asset or services, directed at improving response measures that limit negative consequences, or directed at enhancing recovery operations. The types of information that might be used to describe each alternative might include cost/benefit analyses, insights into any new risks that might be created once the option is exercised, any political risks that might result from the decision, and the feasibility and effectiveness of any proposed countermeasures aimed at risk reduction.

This concludes the discussion on integrated risk management, risk assessment, and alternative analyses. In summary, there is no one set plan or strategy to implementing an integrated risk management system within an organization, nor is there one universal scoring methodology that best addresses all types of risks. Each community comes with its own set of needs and values that can influence the final outcome. Thus, the discipline of risk management has been described more as an art form than a science. What is most important is for the local community risk management planning team and the community's key decision makers to make risk-informed decisions. Led by a capable facilitator, the design and implementation process can be used to develop consensus among the users of the system, ensuring buy-in and ownership of the final design, and perhaps more importantly, buy-in and ownership of the relative risk ranking.

DHS's next step

In November 2011, the Department of Homeland Security released the "National Preparedness Goal" and the "National Preparedness System," two documents based on Presidential Policy Directive 8: National Preparedness. The intent of the directive is to ensure government (local, state, federal, and tribal) and private industry develop certain capabilities needed to address the nation's greatest risks.

Presidential Policy Directive 8: National Preparedness emphasizes that national preparedness is the shared responsibility of the entire community, including all levels of government. The National Preparedness Goal describes the capabilities needed to address five mission areas: prevention, protection, mitigation, response, and recovery. Responsibility for achieving the needed capabilities resides with all community members and not with any single agency or jurisdiction.

The "National Preparedness System" describes the process for achieving the National Preparedness Goal. The first step in that process is the conducing of a Threat and Hazard Identification and Risk Assessment (THIRA). DHS guidance calls for the "whole community" (including federal, state, local, tribal, and territorial governments, the private sector, nongovernmental organizations, faith-based and community-based organizations, and the public) to be included in the process. The rationale for involving the "whole community" rests, in part, on the realization that using the collective resources of all involved parties is likely to minimize costs and avoid redundancy.

As of March 2012, THIRA guidance has not been released. DHS has indicated that communities may use the risk assessment process described by the agency's Hazard Mitigation Program, provided that acts of terrorism are included in the analysis. With this provision, communities would continue to qualify for federal funding.

With the ultimate release of THIRA, DHS will have issued yet another risk scoring methodology. Many of THIRA's predecessors were viewed as overly complex or overly simplistic in leaving too much discretion on how risks were to be scored. THIRA will provide a much-needed benefit to the user community if it is able to achieve the following:

- First, the level of complexity must be balanced with the amount of detail. Details specific as to how to score risk scenarios are needed in order to provide consistency and allow for cross-jurisdiction comparisons. Yet, the methodology cannot be overly complex, as the intended users and local communities have had little formal training in risk management principles.

- Second, the use of THIRA cannot be a one-time effort, as threat levels and hazards are constantly changing, and new risks are likely to be identified in the future.

THIRA would benefit local communities if it provides guidance on how communities can integrate a risk management system into their everyday activities.

- Third, THIRA should be expandable to address risks beyond natural disasters and acts of terrorism. Community leaders and elected officials deal with a variety of risks (safety, financial, legal, etc.), and it would seem prudent that if a community were to allocate limited resources to risk reduction, leaders would want to address all risks to the community.

- Finally, risk scoring methodologies not only need to have sufficient detail as to the data being collected, but need to provide direction if such data is not readily available. This is particularly true in terms of assessing economic impacts or in trying to assess the replacement costs of critical infrastructure in terms of today's dollars.

Conclusion

This article has intended to provide public safety officials and local government officials with information on how to avoid some of the pitfalls with implementing a risk management system, and to offer for consideration an implementation strategy that has been shown to be successful. Based on continuous improvement in performance metrics, documented procedures, and periodic review by top management and key decision makers, communities can integrate risk management principles into everyday work activities across all levels of local government.

Development of a risk scoring methodology that gains the consensus of the end users has been elusive in the past. Integration of a risk management system by a skilled risk manager can lead to many years of greater and better information being provided to the community's decision makers. For elected and appointed government officials, thorough assessment of the upside and downside risks associated with key decisions leads to not only better decision making, but provides a "due diligence" defense if and when disasters occur involving a local community.

MIKE KENEL, Ph.D., has more than 30 years of risk management experience, most notably as an independent consultant retained by state and local governments to develop innovations in emergency management and homeland security practices related to planning, training, and exercises. Formal training in risk analysis and risk management were obtained at the University of Michigan where he received a doctorate degree in toxicology. Project experiences range from assessing the risks and options for addressing an intentional contamination of a municipal water system (USEPA Office of Water) to developing a risk scoring methodology and then applying the methodology at more than 100 critical infrastructure sites as part of an effort to prioritize expenditures (State of Michigan Homeland Protection Board). Additional key experiences include one of the first state-wide food vulnerability assessments (Michigan Department of Agriculture), development of a brand-

protection risk management system for fountain beverages (Coca-Cola Company), development of State Continuity of Government and Operations plans (Michigan State Police), development of online training courses in critical infrastructure protection, continuity of operations, and mutual aid (Water Agency Response Network), and implementation of environmental management systems at more than 20 facilities (Ford Motor Company, General Motors). Most recently, projects supported by Department of Homeland Security and Department of Energy funding include a threat and risk assessment under the National Preparedness System (Franklin County, Ohio) and development of a Local Energy Assurance Outreach Program (Michigan Public Service Commission), respectively.

REFERENCES

DHS Risk Lexicon, Risk Steering Committee, September 2008

Risk Management Practices in the Public and Private Sector: Executive Summary, Office of Risk Management and Analysis, September 2010

Risk Management Fundamentals, Homeland Security Risk Management Doctrine, April 2011

National Preparedness Goal, First Edition, Department of Homeland Security, September 2011

National Preparedness System, Department of Homeland Security November 2011

ADDITIONAL SUGGESTED READINGS

Risk Management: Strengthening the Use of Risk Management Principles in Homeland Security, U.S. General Accounting Office, June 2008.

Strategic Budgeting: Risk Management Principles Can Help DHS Allocate Resources to Highest Priorities, U.S. General Accounting Office, June 2005.

An Integrative Risk Management/Governance Framework for Homeland Security Decision Making, Albert M. Ponenti, Naval Post Graduate School, March 2008.

A Risk Management Standard, The Institute of Risk Management (UK), The Association of Insurance and Risk Managers, 2002.

Chapter 22: A Regional Approach to Public Safety Communications and GIS

BY DAVID SHUEY, CHRISTINE GRIMMELSMAN, ED DADOSKY

Chicago Mayor (former White House Chief of Staff) Rahm Emanuel is credited with saying, "...never let a good crisis go to waste." It was from such insight that Raven911 came into being.

In January 2007, the Cincinnati-Hamilton County Regional Operations Center (ROC) had recently opened, just in time for a severe weather ice storm event that caught the tri-state area off-guard. There were many ROC deficiencies noted during this incident, including the absence of Geographic Information System (GIS) capability. GIS was identified during post-incident analysis as a tool pivotal in support of emergency responders' efforts to better manage disaster consequences of all types.

As a result of the 2007 ice storm, a GIS was installed at the operations center, however, there were limitations. This GIS was difficult to use and included a large amount of functionality that was overly complicated and of limited value to emergency response agencies during a crisis. Frankly, the system had been designed with focus on tax and audit capabilities versus a system intuitive to the first responder community. Furthermore, this GIS was limited geographically to Hamilton County, Ohio, only one of 12 counties in the tri-state region that has come to embody SOSINK (Southwestern Ohio Southeastern Indiana Northern Kentucky).

Throughout 2007, ROC managers continued their efforts with limited success to become proficient with this overly complex GIS that was of little practical value to them. Progress was difficult to gauge. Unbeknownst to SOSINK jurisdictions, in the spring of 2008, they were going to experience their most difficult challenge to date, the breadth of which had never been seen in modern history of the tri-state region, and from one of the most unlikely of sources.

It was a Sunday afternoon on March 14, 2008, and Hurricane Ike had already ravaged much of the Gulf Coast and was destined to become one of the costliest hurricanes in U.S. history.[1] The Ohio Valley looked on secure in the knowledge of a seemingly insurmountable 900-mile buffer between SOSINK and the Gulf Coast, making the region virtually untouchable by even the strongest of storms. Hurricane after-effects, such as extended periods of rain, had been the extent of secondary effects for SOSINK up until this time, but that was all about to change. F1-equivalent tornado-strength winds ripped through the area beginning around 3 p.m. and didn't subside for almost two hours. The winds had come and gone, but the event was just beginning. The 2008 windstorm cut a swath through the SOSINK region, leaving numerous homes damaged and more than 1 million people without electricity.

Calls began to flood the ROC. Requests for basic human needs included food, water, fuel, ice, prescription pharmaceuticals, oxygen, generators, chain saws, debris removal equipment, plywood, hardware, roofing materials, veterinary assistance, and virtually every conceivable request for resources to which modern-day society has become accustomed. The ROC was quickly overwhelmed. One glaring lack of ROC capability was that of resource information, including specific locations of those resources. In other words, the ROC didn't know where anything was.

For the entire seven- to 10-day period, the ROC helplessly reviewed the yellow pages and the limited GIS information available, making it apparent to several personnel trying to fill those resource requests of the need for a GIS resource with emergency responder-centric information as the foundation. Additionally, this newly formulated GIS needed to cover all 12 SOSINK counties versus exclusively Cincinnati-Hamilton County. And finally, the new GIS configuration had to be so simple that "even a caveman could use it." This was a very daunting task for a bunch of firefighters and emergency planners, but nonetheless, this was the task at hand.

Cincinnati-Hamilton County, as a Department of Homeland Security Urban Area Security Initiative (UASI) grant recipient, took the lead in the interests of the SOSINK region and revitalized a dormant relationship with the Ohio-Kentucky-Indiana Regional Council of Governments (OKI) GIS Division. Discussions began in earnest geared toward the development of an emergency response-centric GIS. Emphasis would be placed on Critical Infrastructure Key Resources important to emergency responders and sufficiently descriptive, ultimately providing responders access to the types of information critical to the management of incidents like the Hurricane Ike windstorm. The overriding focus was to craft this GIS such that it was sufficiently detailed but simple to use. That description is befitting of Raven911 (formerly ROGREMS-**R**esponder **O**perations **G**IS **R**egional **E**mergency **M**apping **S**ystem). Because the name ROGREMS proved to be somewhat awkward and difficult to remember, the system name was changed to Raven911. The Raven is a notably intelligent and agile bird[2] that has become synonymous with this new GIS approach, and the 911 is representative of the emergency response focus of the system.

The emergency response community clearly values such a system for rationale already provided. OKI has benefitted by strengthening emergency responders' capacity to mitigate transportation incidents whether on expressways, bridges, or related transportation modalities. Raven911 also afforded OKI GIS an opportunity to begin an amalgamation of GIS information heretofore stove-piped throughout the eight counties that comprise the OKI region. This was a unique opportunity in the interests of the OKI counties to coordinate development of this comprehensive but straightforward GIS map system that not only aids emergency responders, but also helps the region in pursuit of a coordinated transportation strategy and commensurate grant funding...nothing like a good crisis to help pull off such a feat.

While it is correct that the primary data sets identified for Raven911 came from Hurricane Ike, other important GIS information was researched through catastrophic disasters such as 9/11 and Hurricane Katrina. Notably, the types of resources and infrastructure information identified by responders from Ike, Katrina, and 9/11 were similar enough that a responder-centric GIS could be formulated with a reasonable degree of confidence to have value added for most emergencies, including floods, tornadoes, earthquakes, hurricanes, chemical incidents, or acts of terrorism. The following text will detail how this information is validated by emergency responders and also describe how OKI GIS experts continue to identify useful tools to integrate into Raven911, further enhancing system functionality.

Background

With a regional population of close to 2.1 million individuals and an impressive 4,686 square miles, the Cincinnati Urban Area Security Initiative (UASI) and the Cincinnati Metropolitan Medical Response System (MMRS), a division of the Cincinnati Fire Department's Homeland Security Unit, are charged with the daunting tasks of mitigating, planning and preparing for, responding to, and recovering from large-scale natural or manmade disasters occurring within their defined planning region. The Cincinnati urban area's emergency planning region is comprised of 12 counties and three states encompassing Adams, Brown, Butler, Clermont, Clinton, Hamilton (City of Cincinnati), Highland, and Warren counties in Ohio; Boone, Campbell, and Kenton counties in Kentucky; and Dearborn County in Indiana (Figure 1). The SOSINK planning region is highly unique in that it is the largest of its kind in any of its three counterpart states, it is one of less than a handful of multistate emergency planning regions anywhere in the

Figure 1. Depiction of the SOSINK Emergency Planning Region.

United States, and finally, it has successfully formed a true regional response capability irrespective of conflicting state and local legislation — not to mention its ability to create a common operating picture among disparate jurisdictional, agency, disciplinary, and geographic boundaries.

With funding provided primarily through a cadre of financial programs under the umbrella of the U.S. Department of Homeland Security Department's Homeland Security Grant Program, emergency planning funds are subsequently passed through the Ohio Emergency Management Agency and eventually funneled to a regional steering committee via two primary points of contact: Edward J. Dadosky of the Cincinnati Fire Department and Mike Snowden of the Hamilton County Emergency Management Agency. These points of contact were consequently selected as leaders in the collaborative effort largely because they represent two of the most critical emergency response agencies within the single largest population in the region: Cincinnati and Hamilton County. While the primary points of contact serve as a mechanism for communication, administration, and oversight of financial aspects of the SOSINK initiative, the responsibility for funding, strategic planning, and project-related decisions falls squarely with the individuals who make up the SOSINK steering committee.

The steering committee is comprised of 35 members representing each of the participating SOSINK counties. And, while the whole of the decision-making process is accomplished at the steering committee level, its members do so with guidance provided by specialized subcommittees representing each of 12 emergency response disciplines, including fire, emergency medical services, hazardous materials, law enforcement, hospitals, public health, public works, volunteer organizations, state/local and federal government, private industry, communications, and emergency management.

Early discussions surrounding the creation of a regional emergency mapping system were summarily dismissed based on a variety of rationale. Central to early objections were seemingly insurmountable obstacles and questions such as:

- What would the benefit be?

- Who (person, agency or group) would take the lead in developing and maintaining such a system?

- How much would this regional mapping system cost to create and maintain, and how would it continue to be funded?

The windstorm of 2008 put all those concerns and questions to rest (Figure 2). A primary lesson learned from this multi-day natural disaster was that the region lacked not only an interactive geographic representation of itself, but also immediate access to its assets, critical infrastructure, and resources. In essence, this unprecedented storm revealed gaps that quite simply could no longer be ignored.

Driven by the experiences in the Emergency Operations Center during this devastating emergency, the Cincinnati Fire Department's Homeland Security Unit readily accepted the challenge of creating a regional mapping system. While members of the Fire Department's Homeland Security Unit undoubtedly possessed the emergency response perspective and experience required to complete the task, it was not without noting that they lacked some of most critical aspects of the projects, specifically, technical (GIS) expertise, a methodology for data collection, financial support, and the manpower required to undertake such as massive project. Undeterred as many firefighters are, the unit's leadership began seeking the necessary resources to accomplish what would turn out to be not only one of the region's most monumental undertakings, but also one if its most successful.

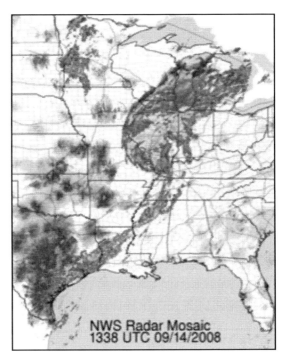

Figure 2. 2008 Windstorm radar snapshot.
(COURTESY: NATIONAL WEATHER SERVICE)

Overcoming obstacles

For perhaps the most critical aspect of the mapping project, the technical/GIS piece, a partnership was sought with the existing regional planning authority, OKI (the Ohio-Kentucky-Indiana Regional Council of Governments). For its part, OKI was equally in favor of a mapping partnership and eager to contribute a portion of its funding, as well as all of the software, equipment, personnel, and baseline data that ultimately paved the way for the RAVEN911 Mapping Project.

When it came to the data collection component of the project, the Homeland Security Unit's counterpart, the Cincinnati Metropolitan Medical Response System (MMRS), was the logical choice. The decision to utilize the MMRS planner to manage the data collection function was neither impulsive nor random, as Cincinnati MMRS already had established a method for collecting, storing, and maintaining data, and it recently had compiled and published a successful regional asset inventory of all fire and emergency medical services departments, emergency response equipment assets, and personnel within the 12-county, three-state SOSINK region.

When it came to funding, support for the project was gleaned from the SOSINK subcommittees, and an appeal for monetary support was subsequently presented to the SOSINK steering committee. Ultimately, the steering committee agreed not only on the importance of the project but also the idea of leveraging Federal Highway Funds acquired by OKI in conjunction with UASI funding secured from the U.S. Department of Homeland Security to get the project under way.

There was one final obstacle: the lack of manpower. To overcome this problem, Cincinnati Fire Department's administration was approached with the idea of utilizing light duty fire personnel to assist with data collection activities. Light duty (or limited duty) fire personnel are sworn, uniformed members of the fire department who suffered physical disabilities, injuries, illnesses, or conditions that were not significant enough to prevent their employment, but significant enough to prevent them from undertaking their normal firefighting responsibilities for an established amount of time. Traditionally, light duty personnel are assigned to assist with administrative tasks, i.e., delivering mail, assisting with special short-term projects, and a seemingly endless number of fire department functions that must be accomplished but have no fire/civilian permanently assigned to such tasks. For its part, the Cincinnati Fire Department viewed the use of light duty personnel for this purpose as both unique and prudent. In retrospect, this particular decision proved to be one of the most pivotal, and it unequivocally contributed to the success of the RAVEN911 Project.

Administrative structure

Once the key responsibilities of the project were defined, a simple role-based organizational structure emerged (Figure 3). Individual roles within the structure included:

SOSINK Steering/Subcommittee—Provides continued authorization and funding in support of the RAVEN 911 Mapping System. Assists in developing critical infrastructure/asset layer list to be included within the regional map.

Project Administrator—Provides guidance, direction, and support to RAVEN 911 GIS / Technical and Project Development and Data Collection Managers. All major decisions regarding the mapping project are presented for approval by the Project Administrator. The Project Administrator determines the next steps (i.e., end-user training, program or capability enhancements). This individual also communicates future goals and shares project successes with SOSINK steering/subcommittee members.

GIS/Technical Manager—Responsible for all technical developments (widgets). Provides oversight of final data verification, geolocation, and import processes. Recommends and orchestrates program enhancements (where possible). Seeks latest technology and offers ideas and support for program utilization.

Project Development/Data Collection Manager—Oversees data collection efforts. Manages primary data collection verification process. Maintains storage of and updates data already on file. Recommends enhancements to tools and capabilities. Provides demonstrations to showcase system capabilities. Authorizes end-user applicants. Serves as liaison to communicate end-user feedback to GIS/Technical manager. Utilizes system during significant events to capture and communicate data.

End Users—Utilize system for planning, training, exercise, and emergency purposes. Provide feedback and recommendations for new or enhanced tools, and communicate these to Project Development/Data Collection Manager. Collect and/or verify local data as requested.

As for data collection itself, the process can be long and arduous. Some preparatory steps taken by the development team established a baseline for the mapping system and ultimately proved to be a worthwhile effort. That said, it is prudent for data managers to anticipate a significant amount of preliminary research prior to delving into data collection activities in earnest. In the case of the RAVEN 911 Mapping System, an extensive amount of preliminary planning and a variety of research modalities were employed to help define the data that would be collected and how the data collection process itself would proceed.

One of the first steps taken in planning for the data collection process was the establishment of a few foundational principles related to ALL data collection and management activities. These included:

1. Data collection—specifically the type of data that would ultimately be captured—would be driven by the end user: the emergency responder.

2. Data accuracy shall be upheld as the most important component of the mapping project. Without exception, inaccuracies in data collection constituted avoidable dangers to emergency responders, the public at large, and the project overall. As such, a process for accurate data collection, and geospatial and/or visual verification of data (where possible) was devised.

3. Context is everything.

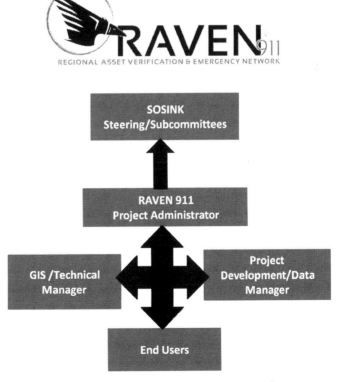

Figure 3. RAVEN911 mapping system role based organizational chart.

Once these founding principles were defined and agreed upon by the development team (Project Administrator, Technical/GIS Manager, Project Development/Data Collection Manager), they established guidance for the remainder of the data collection effort. In fact, these foundational principles continue to guide data collection and integration of data into the mapping system even in the present day.

Existing mapping systems

Among the first research activities undertaken was a comprehensive review of existing emergency mapping systems. At the onset of the mapping project in 2008, this task was far less time-intensive than could be expected in the present day because very few emergency mapping systems either existed or were readily accessible at that time. With respect to what was learned, this effort revealed that most emergency mapping systems were singularly state-focused (rather than multi-state, or regional) in their scope, and seemed at least anecdotally to focus on emergency management functions (such as automating the damage estimation process) or were geared specifically to functions such as pre-planning for large-scale emergencies or disasters. This in no way suggests that these systems are in any less valid or useful; rather, they simply did not reflect the vision of the RAVEN911 development team. This particular step in the research process yielded a great deal of background information and helped in formulating a design and sharpening the focus of what ultimately would become the RAVEN911 Mapping System.

Learning from past experience

Another research effort undertaken to help define the data that would be most useful was the solicitation of advice from subject matter experts. For the purpose of this project, subject matter experts were selected based on their role(s) in a large-scale emergency or disaster. Among the experts interviewed was local Fire Chief Paul Wright of the City of Montgomery (Ohio) Fire Department. On April 9, 1999, the City of Montgomery (among others) suffered a devastating tornado. As the incident commander, Chief Wright's insights were exceptionally enlightening to the development team in determining what types of data should be collected for the mapping system. Similarly, individuals who operated within the Cincinnati-Hamilton County Regional Emergency Operations Center during the windstorm disaster of 2008 were polled to get a feel for the types of resources or infrastructure that was deemed critical during a widespread, multi-day emergency. At national conferences, information was gathered from those who played a pivotal response or emergency management role in such devastating emergencies as the attack on the World Trade Center and Hurricane Katrina to identify resource priorities. Finally, the expertise of our State Urban Search and Rescue Team was leveraged to get their opinions on the types of information that would be most useful to have on hand during various stages of a disaster (from deployment to demobilization).

Utilizing this feedback, the RAVEN911 development team noted some generalities and utilized them to generate an initial (DRAFT) list of resource layers to be incorporated into the RAVEN911 Mapping System. This initial list of layers was subsequently presented to the regional, multi-disciplinary steering committee for their feedback. In the end, it was the Regional Steering Committee members who identified the initial list of data layers to be included in the mapping system.

Not surprisingly, the initial resource "wish list" was simply too substantial to be practical in the context of an electronic mapping system. As such, the list was subsequently subdivided into "map layers" and "other" resources (backhoes, water movers, debris clean-up services). This is not to say that the layers relegated to the "other resources" category were ignored. In fact, the list of "other resources" was assigned to a resource coordinator who collected and verified the data in a similar manner but stored it in a separate resource database. As for the "mapping layer," it continues to be dynamic and ever-changing based on user feedback and lessons learned from planned events, exercises, trainings, and actual emergency incidents.

As for prioritizing which data should be collected first, a "low hanging fruit" approach was taken—that is, seeking existing resource or asset lists that were likely to be complete, accurate, and readily accessible. In fact, the first layer of data integrated into the mapping system was derived internally through the Cincinnati Metropolitan Medical Response System's regional fire and EMS asset inventory, which had been completed only months before. From there, attention was given to existing resources or registries from more typical sources, such as the phone book and the Internet. In some cases, data was captured through state and federal registries, and queries of end users. Where absolutely required, Freedom of Information Act (FOIA) requests were submitted. Of all research collection modalities, the FOIA method proved the least successful and yielded rejections citing a "risk to homeland security" as the rationale.

Beyond FOIA requests, obstacles were encountered throughout the data collection process. Among the layers most difficult to collect were chemical facilities and utilities, such as electrical substations, gas storage locations, cell phone towers, and water/hydrant system data. As encumbrances were encountered, researchers simply diverted their efforts to more achievable data collection efforts. In some cases, the more difficult data layers were assigned to interns at OKI who successfully applied more technological modalities to collect data, such as visual scanning of the region using Pictometry and LiDAR data. While this last method was certainly the most time-intensive, it proved exceptionally successful and, in at least one case, yielded several thousand critical infrastructures (cell phone towers) where less than 100 existed from all previous data providers combined.

Light duty personnel

Where practical, light duty firefighting personnel worked with the data collection manager to make direct contact (via phone, fax, email, or in-person interviews) with facilities such as nursing homes, colleges, law enforcement agencies, etc. Such solicitation was deemed essential to assist emergency responders in quickly and accurately identifying features such as occupancy, emergency contact numbers, etc., during actual emergencies. Similarly, light duty personnel were utilized externally to capture and verify tangible/visible critical infrastructure from its exact location. In such cases, the Cincinnati Fire Department (and on one occasion, the Covington Fire Department) light duty personnel employed the use of GPS equipment, photography, and note taking to collect the requested data.

One example of field data collected by light duty firefighters was river flooding locations. For this layer, firefighters were paired (for safety) and were instructed to stand at the edge of the Ohio River during actual flooding conditions. There, they captured GPS coordinates to show the actual extent of flooding. In another example, light duty personnel (again paired for safety) collected data on nearly 1,500 at-grade railroad crossings. In the case of the railroad crossing data collection effort, light duty firefighters were asked to stand in the center of every at-grade rail crossing in the region. There, they captured exact coordinates via a GPS device, photographed Department of Transportation identification signs (less than half were present), and documented any infrastructure of relevance in the immediate vicinity (schools, chemical facilities, etc.) Again, in keeping with our pledge to ensure accuracy, all data was verified visually and/or geospatially, prior to being imported into the RAVEN911 Mapping System. At the time of this writing, data layers included Fire/EMS stations, hospitals, law enforcement agencies, water rescue assets, mass decontamination trailers, mass casualty trailers, schools, chemical facilities, colleges and universities, gas and electric infrastructure, among others.

Application framework

The RAVEN911 application is based upon ESRI's ArcGIS Viewer for Flex. The system was built using ArcGIS 10.0, including ArcGIS Server and ArcGIS Desktop software. It is a web-based rich Internet application requiring the user to have a connection to the Internet and Adobe's Flash Player plug-in for the browser of their choice. The application consumes ArcGIS services, including mapping, network analysis, geocoding, geometry, and geoprocessing services, utilizing the Flex framework.

Overview of functionality

RAVEN911 has the ability to overlay numerous critical infrastructure layers—including chemical facilities, daycare centers, fire stations, hospitals, nursing homes, power plants and utility infrastructure, Red Cross shelters and schools—onto multiple base maps. It also has the ability to display real-time weather data from a network of weather stations across the greater Cincinnati region, a weather radar loop from the past hour, USGS feeds for stream gauges and earthquakes, and the location of active wildfires. A network of more than 100 traffic cameras provided by the Advanced Regional Traffic Interactive Management & Information System (ARTIMIS) for Greater Cincinnati allows emergency responders to monitor conditions along interstates in real time. A containment zone widget allows users to define a zone of containment. The application then calculates road closures to the nearest address, allowing authorities to rapidly isolate the incident zone. A drive time calculator is included and can calculate up to three drive time polygons concurrently. This allows emergency responders the ability to calculate response time to a given incident from any location in the region. Also included is a driving directions tool that will generate directions to any address in the region. An Emergency Response Guidebook widget allows emergency responders to create an isolation zone for a chemical release or spill. The widget accounts for the type of material, spill size, time of day, and current weather conditions. Once the isolation zone is determined, road closures are automatically calculated and containment can begin. A Plume widget allows the end user to load chemical plumes from various sources. RAVEN911 has a built-in U.S. National Grid tool that can be utilized in search and rescue missions. It also has a Twitter search widget, which allows users to search tweets based upon their spatial location.

Achieving goals

With respect to its preliminary goals, the RAVEN911 Mapping System has been activated and utilized for planned events, strategic planning, and emergencies. As for planned events, RAVEN911 was used to prepare for and monitor Riverfest (one of Cincinnati's largest annual events with some 500,000 persons in attendance). RAVEN911 is further being utilized to prepare for unique and highly anticipated special events such as the 2012 World Choir Games.

Outside the Cincinnati area, the RAVEN911 system was deployed to capture data during the Indianapolis State Fair Stage Collapse (2011) and the Chardon High School Shootings (2012). In the wake of these significant events, data collected utilizing RAVEN911 tools were compiled into comprehensive report formats and submitted (by request) to the Indiana State Police and Federal Bureau of Investigation (Columbus FBI office via supervisory special agent Robert White II, on behalf of the Cleveland FBI office). Regionally, RAVEN 911 was utilized to assist with emergencies such as a party boat collision on the Ohio River, an explosion/arson incident, the collapse of a Cincinnati casino under construction and the devastating tornadoes that decimated parts of two SOSINK regional communities and killed four people in the spring of 2012.

Due to its popularity, requests for access to the RAVEN911 system have been received from agencies and emergency responders from across the country. Additionally, access to the RAVEN911 Mapping System has been granted to a host of federal agencies, including the Department of Homeland Security, (TSA/Air Marshalls/Intelligence sections), the Federal Bureau of Investigation, the Food and Drug Administration, the Environmental Protection Agency, the Internal Revenue Service, the Department of Energy, and the U.S. State Department.

DAVID T. SHUEY, GISP, is the GIS Division manager with the Ohio-Kentucky-Indiana Regional Council of Governments in Cincinnati. He is responsible for managing the implementation, use and operation of GIS software and databases for the council's eight-county region. Since 2005, David has served as an adjunct professor in GIS at the University of Cincinnati where he conducts several GIS courses. David holds a B.A. in Geography and a master's degree in Community Planning, both from the University of Cincinnati.

CHRISTINE GRIMMELSMAN brings 20 years of experience in the fire and emergency medical services fields to her current role as an emergency planner with the Cincinnati Metropolitan Medical Response System. In her present assignment within the Cincinnati Fire Department Homeland Security Unit, she serves as a liaison with all 12 first responder disciplines (which include fire, EMS, law enforcement, emergency management, state, local and federal governments, hospitals, and public health services, among others). In her present position, Christine is charged with ensuring that the Southwest Ohio, Southeast Indiana, and Northern Kentucky planning region is prepared to respond in the event of a large-scale emergency or disaster. Chris presently maintains certifications as a firefighter, paramedic, emergency medical services instructor, and hazardous materials technician, and holds a master's degree in Public Administration from the Northern Kentucky University. Chris continues to work as a firefighter/paramedic on a part-time basis with the Golf Manor Fire Department in Hamilton County and Felicity Franklin Emergency Medical Services in Clermont County.

District Chief ED DADOSKY is a 28-year veteran of the Cincinnati Fire Department. He was certified as a paramedic in 1995, was promoted to District Fire Chief in 1999, and was promoted to Assistant Fire Chief in 2011. Chief Dadosky graduated from the Cincinnati Police Academy and became a sworn Police Officer in 2001. Chief Dadosky currently serves as the Assistant Fire Chief of the Administrative Division for the Cincinnati Fire Department, Department of Homeland Security Cincinnati Urban Area Security Initiative (UASI) Core City Point of Contact (POC), and as the regional Metropolitan Medical Response System (MMRS) Coordinator. His responsibilities for the fire department include logistics, supply, fleet management, facilities management, and information technologies inclusive of the tri-state regional operations center. Chief Dadosky holds a bachelor's degree in Business Administration from the University of Cincinnati and a master's degree from the Naval Postgraduate School located in Monterey, California.

ENDNOTES

1. http://en.wikipedia.org/wiki/Hurricane_Ike

2. http://animals.nationalgeographic.com/animals/birds/raven/

Chapter 23: Continuity of Operations (COOP) Automation Program

BY ROBERT JONES

The Montgomery County, Maryland, Office of Emergency Management and Homeland Security (OEMHS) develops and administers implementation of the Continuity of Operations (COOP) Plan cooperatively with county departments, agencies, and municipalities. With approximately 40 departments, 15 municipalities, and 6 agencies, most with a large number of subordinate organizations and buildings, it is imperative that the county government develop a continuity of operations program that ensures all organizations are prepared to provide mission-essential functions during times of disaster.

OEMHS, in close collaboration with the National Capital Region WebEOC administrative support team, has developed a COOP program management tool that is integrated with the county's WebEOC emergency management system. The WebEOC system is a web-enabled crisis information management system widely used by various emergency response and management personnel across the country. The COOP automation tool simplifies the county organization's COOP plan development process and optimizes the management of resources, such as alternate office space and facilities, resources and supplies, and communications capabilities. It captures data that allows for the analysis needed to ensure that COOP may be executed without disrupting another organization's efforts.

Montgomery County is one of the few jurisdictions in the United States that has successfully merged the Emergency Operations Center (EOC) emergency management with COOP operations into one system using the WebEOC system. Situational awareness, leadership "dashboard" reporting, collaborative information exchange, timely and orderly planning support, and ease of use has greatly increased the readiness of the county organizations during times of emergency.

Problem

The Office of Emergency Management and Homeland Security (OEMHS) is tasked to lead and manage the countywide COOP program to support numerous, diverse departments, agencies, and municipalities. Federal, state, and county-level guidance—such as Federal Preparedness Circular (FPC) 65, State of Maryland Emergency Management Agency (MEMA) directives, and Montgomery County legislative requirements—were used to establish a COOP program.

A viable COOP program must be maintained at a high level of readiness, be capable of implementation with and without warning, be operational within 12 hours after an emergency, be able to maintain sustained operations for at least 14 days, and take maximum advantage of existing organizational field infrastructures.

Initially, most county organizations did not have a COOP plan, or had poorly designed and outdated plans. There was limited experience in developing plans, and no training and exercise program for COOP operations. There was very limited or no COOP collaboration with other county departments and agencies. The Emergency Operations Center (EOC) staff members, and EOC system processes and programs were not being used to support COOP operations in the county.

Because the majority of COOP activation needs take place during a major emergency or disaster, the county had to simplify planning and preparedness, and integrate COOP activations and operations with EOC response and recovery efforts. Integrating COOP into WebEOC presented a cost-effective solution that reduces the cost of planning and operations, and leverages existing expertise already found in the emergency operations and response network.

Response

Description of the program

The implementation team identified the following key objectives to be addressed for the county's COOP program:

- Minimize injury, loss of life, and property damage;

- Ensure continuous performance of mission-essential functions and operations;

- Protect facilities, equipment, records, and other assets;

- Reduce or mitigate disruptions to operations;

- Identify and designate principals and essential support staff—COOP program managers/planners and Emergency Response Group (ERG) personnel;

- Facilitate decision-making for execution of the plans and the subsequent conduct of operations;

- Achieve a timely and orderly recovery from an emergency and resumption of full service to Montgomery County residents;

- Integrate the COOP program into the EOC planning, response, and recovery operations effort; and

- Develop a convenient, secure, reliable, scalable, fast, and cost-effective web-based solution that will enhance the capabilities of our COOP and EOC staff members.

OEMHS, in partnership with the National Capital Region WebEOC administrative support team, developed an automated COOP program management tool that is integrated with the county's WebEOC web-enabled crisis information management system, which is used by various emergency response and management personnel. The automated COOP tool provides multiple templates for creation of COOP plans; enhances the EOC staff's management of resources, such as alternate office space and facilities, resources and supplies, and communications capabilities; and captures COOP data that allows for the analysis of information and ensures the organization's COOP plans are executable without disrupting another organization's efforts.

The WebEOC tool automates the COOP planning, response, and recovery effort, and is critical to the overall success of the county effort to provide mission-essential functions and services during times of disaster and extreme emergencies for the residents of Montgomery County, Maryland.

The WebEOC template board views were created and customized to address the various key components of COOP program development and documentation. They all reside within the WebEOC information management system that is utilized during activation of the EOC. The templates are easy to use, maintain, and update. The tool allows the OEMHS and EOC staff to monitor and manage COOP activities of multiple departments proactively, and prior to any activation it can be used to resolve multiple usage conflicts and issues that may occur between organizations. Additionally, "dashboard" reporting provides reports for leadership to effectively manage resources to support mission-essential functions.

The COOP board enables different departments to independently and securely initialize their COOP plans, and to submit and manage COOP information for their discrete units, while enabling a higher level supervisory position access to view all COOP information. Access to view stored information is rigidly controlled. The COOP board within WebEOC uses multiple views and nine (9) major input views for recording various COOP components. All board views can be printed in WebEOC and collated to form the printed COOP plan as a back-up in case Internet access is not available for any reason.

User interface—COOP PM Control Panel

The following screen (see next page) depicts a list view that shows the COOP Department Control Panel where each organizational unit developing a plan will have a separate Department Control Panel with buttons on the left to access the identified data. Descriptions are on the right. Each button opens a display view, and each display view has an associated input view as appropriate.

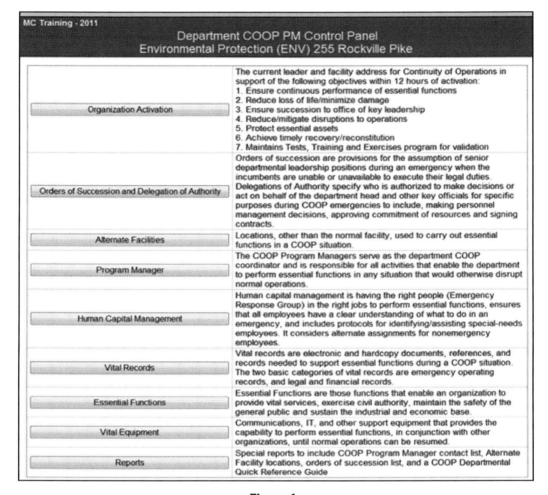

Figure 1.

COOP Organization Activation—This component is for specifying who will be the leader for the department and where the current facility is located. Each field provides a date/time field to indicate when the information was last updated.

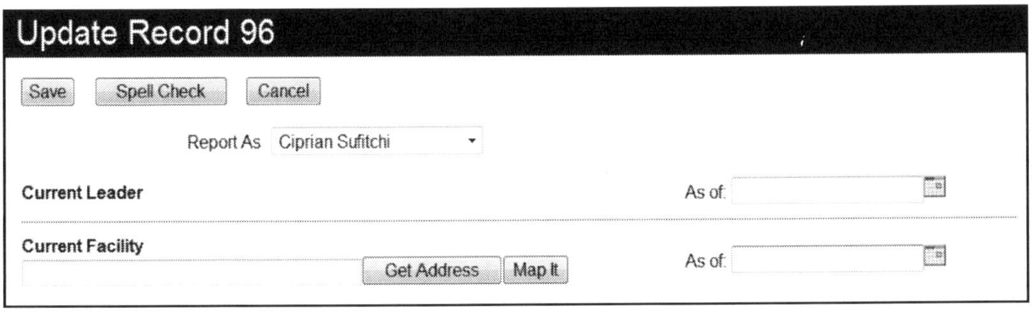

Figure 2.

COOP Order of Succession and Delegation of Authority—This component displays the individuals and supporting information identified for the order of succession as well as delegation of authority information.

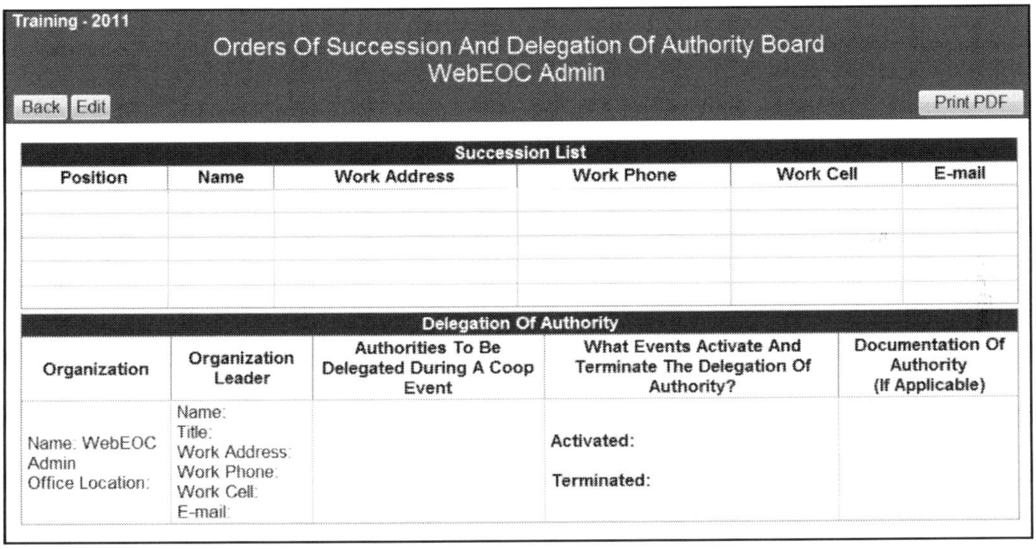

Figure 3.

COOP Alternate Facilities—This list displays the location, address and potential phone numbers for alternative facility locations.

254 CHAPTER 23: CONTINUITY OF OPERATIONS AUTOMATION PROGRAM

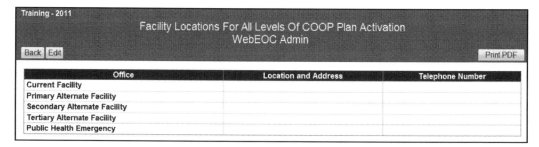

Figure 4.

COOP Program Manager—This list view displays the COOP program managers that will serve as the department COOP coordinators. In addition, the organizational *mission* and *special considerations and/or responsibilities* can be annotated in this board view.

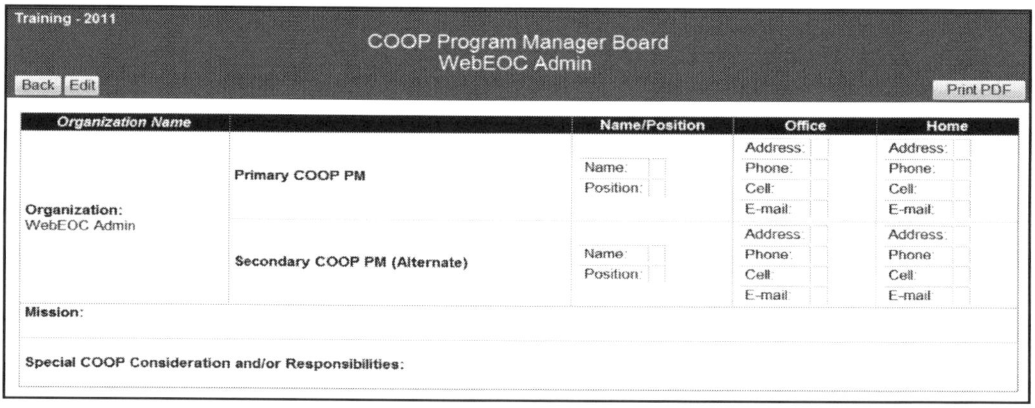

Figure 5.

COOP Human Capital Management—This list view provides an area to view all employees identified to support essential department functions.

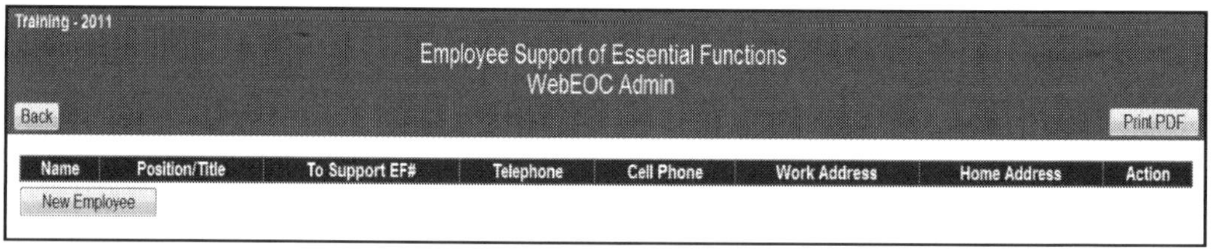

Figure 6.

COOP Vital Records—This list view displays documents that are considered vital to a department during a COOP situation.

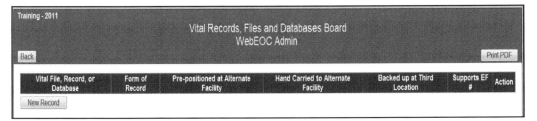

Figure 7.

COOP Essential Functions—This list view displays the essential functions to enable the department to provide vital services.

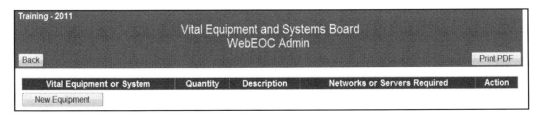

Figure 8.

Vital Equipment—This list view displays all necessary equipment needed to perform essential functions.

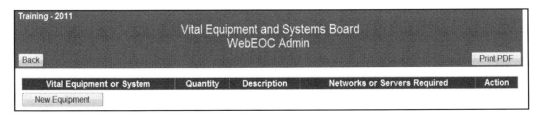

Figure 9.

COOP Reports—This selection offers an area to select from four different reports.

COOP Quick Reference Guide	This Quick Reference Guide is designed to assist the EOC Emergency Management Group members with providing support to various departmental Continuity of Operations (COOP) activations and alternate facility deployments. Use this to view, analyze, and report on departmental COOP plans and operations.
Dept. COOP PMs	This report provides a complete list of all assigned Continuity of Operations (COOP) Program Managers (PM) who are responsible for COOP planning, preparedness and activation of their COOP plans on behalf of their departmental leadership. Use this for COOP PM contact information.
Dept. COOP Alternate Facilities	This report provides a complete list of all potential alternate facilities that may be used in the event departments need to execute their COOP plans. Use this alternate facility list to aid departments in their deployment efforts and provide proactive support if there are multiple organizations requesting to use the same alternate facility.
Dept. COOP Succession Plan	This report consolidates all departmental succession lists for departments that have essential functions and require COOP plans. Use this orders of succession list to aid departments in their efforts to maintain essential functions.

Figure 10.

County departments and agencies use the COOP board list and display views to independently submit and manage information, while enabling a higher level supervisory position to view all COOP information for the county. For example, the COOP Organization Activation information noted above is completed by each organization. This information is consolidated in the County WebEOC Controller Panel displayed in the screen that follows.

User interface—Consolidated County COOP Organization Activation Control Panel

The WebEOC controller and EOC OEMHS staff can monitor and support county departments and agencies as they execute their COOP plans. This "dashboard" (see Figure 11) can, at a glance, depict who is in charge and where any given department and agency is located during a disaster or significant emergency. This board view demonstrates the integration of COOP within the WebEOC application, where, during activations of the EOC and activation of COOP responses, authorized personnel have centralized access to a wide range of information and situational awareness to coordinate the response efforts.

The cost of the program

The county is using the WebEOC system that already has been purchased to support the emergency response and recovery efforts in the EOC. This WebEOC hosted solution includes maintenance and template board development support by the NCR WebEOC administrative support team. This team's costs are included in the WebEOC license. There was no additional cost to the county to develop or maintain the COOP automation tool in WebEOC. The design and development effort was achieved by three people over a four-month period from January to April 2011. No additional costs were incurred to develop this solution.

Figure 11.

Results

Since its initial launch in May 2011, the COOP automated tool has been used by sixty (60) county organizations to develop their organizational COOP plans. In the first quarter of 2012, we have added an additional twenty (20) organizations and expect to add additional organizations throughout the year. As a result, the county now has departmental succession plans that can be updated easily every month, clearly defined delegation of authority for multiple departments, detailed departmental mission-essential functions, appropriately identified organizational alternate facilities, interoperable communications, properly identified vital records and databases, and a program to effectively test, train, and exercise our county COOP program.

Additionally, the county organizations now have the ability to maintain situational awareness throughout any emergency or disaster by simply accessing WebEOC, which is used by the Emergency Management Group who support EOC activations, response, and recovery operations; and the COOP program managers and the COOP emergency response group members who support COOP activations. The county has successfully merged the EOC emergency response efforts with COOP operations. This allows for a coordinated emergency management effort to address the disaster and integrate the departments' COOP efforts resulting from the disaster.

The county leadership now has an integrated emergency management and COOP solution that ensures a common operating picture, updated situational awareness, and an effective situational reporting process.

The COOP solution developed in Montgomery County, Maryland, has been requested by and shared with multiple organizations around the United States, including the Transportation Security Administration (TSA), Harvard University, Camp Lejeune Marine Corps, Fort Lee Army Base, and multiple county and state emergency management organizations in Oregon, Washington, California, Texas, North Carolina, Virginia, Massachusetts, and Oklahoma.

The COOP automation solution in WebEOC provides Montgomery County COOP program managers, COOP Emergency Response Group members, and EOC members with an online tool to prepare for, respond to, and recover from an emergency or disaster. Montgomery County is now one of the few government agencies in the United States who has successfully merged into one system EOC operations with COOP operations using the WebEOC system. Situational awareness, leadership "dashboard" reporting, collaborative information exchange, timely and orderly planning support, and ease of use has greatly increased the readiness of the county organizations during times of emergency. The county residents are better served as a result of the county efforts to integrate emergency operations with continuity of operations into one automated system.

Finally, by collaborating development with the NCR WebEOC team, and using existing expert staff already performing emergency management functions, there was no additional cost of development for this COOP system application. This tool is invaluable. While it is difficult to quantify savings that are realized due to the ease and efficiency with which the county can coordinate and deliver critical public safety services during emergencies, the fast and effective deployment of coordinated resources helps save lives, preserve property, and minimize the hardships of emergencies. In summary, this tool significantly helps government continue operating when all else is shutting down.

By developing our own solution, we saved a minimum of $210,000 because we did not have to purchase and maintain an off-the-shelf vendor product. Sharing this solution with other organizations could result in very cost-effective and efficient solutions to meet the need of integrating COOP and emergency management operations.

Future developments

The COOP system tool will continue to undergo revisions to enhance the various organizational COOP program managers' planning efforts. This tool will grow with the needs of the people who are responsible for continuity operations. As COOP program managers learn more and become better trained, the COOP system tool will become more robust to meet their needs. Future enhancements include a communications systems view, a pre-surveyed alternate facility availability view, vulnerability and risk view, continuity agreements view, and other continuity-related views that are important in the planning process.

This COOP tool has significantly enhanced our county departmental readiness during times of emergency, but it is just one piece of the overall COOP program. It is the planning piece or just one leg of a three-legged COOP program stool. It is an important leg to the stool and in many ways is sufficient for a number of organizations in our countywide COOP program. However, to really take it to the next level and stabilize the stool, Montgomery County is building two additional legs of the stool: a COOP Job Aid Board and a COOP Enhancement Suggestions Board. These two additional boards (legs) will complete the COOP program stool and create a program that is not only well planned, but easier to implement and execute in an emergency. We will have a planning leg, implementation and execution leg, and an enhancement and improvement feedback leg.

The COOP Job Aid Board is designed to consolidate the various COOP plans, guidelines, and checklists. These documents are intended to support the preparedness planning, operational processes and implementation actions required to prepare for and execute a COOP plan. The board is organized into the four major phases of continuity: readiness and preparedness, activation and relocation, continuity operations, and reconstitution. The various support documents are listed in the appropriate continuity phase and are attached as a Word, Excel, HTML, or other document. In addition, once added, a revision date will be loaded into the board. Organizations will have one year to update the document either by testing, training, exercising (TT&E), or reviewing the document. The board's due date will turn yellow when it is due for review in 90 days and will turn red when the board is past due. A master control board will capture the yellow and red due dates, and will help the county COOP program managers to contact and remind the organization to update and review their documents. The table on the following page is a draft of what the new board will look like.

The plans, guidelines, and checklists listed above will enable organizational COOP program managers, Emergency Response Groups, and other COOP-related personnel to implement and execute the COOP plan. These documents will be enhanced and improved upon as we all gain experience and utilize the documents as a part of our test, training, and exercise program. Additional plans, guidelines, and checklists will be created in the future and integrated into the COOP Job Aid Board. Note: For federal, state, and other organizations that have the capability, this board could be modified to include a devolution phase with the appropriate plans, guidelines, and checklists for devolution.

CHAPTER 23: CONTINUITY OF OPERATIONS AUTOMATION PROGRAM

COOP Job Aid Board			
Plans, Guidelines, or Checklists	Revision Date	TT&E Due Date	Document
Readiness and Preparedness			
COOP Plan Handbook			
Business Process Analysis (BPA)			
Business Impact Analysis (BIA)			
COOP Program Manager Checklist	01/12/10	12/05/12	Attachment
Family Support and Preparedness Plan			
Building Evacuation Plan	03/23/09	05/10/12	Attachment
Facility Vulnerability Assessment			
Activation and Relocation			
COOP Alert and Notification Checklist	08/30/22	03/10/12	Attachment
Employee Advisories, Alerts and Instructions			
Personnel Accountability Guideline			
Initial Actions and Advance Team Checklist			
Alternate Facility Deployment Checklist			
COOP Activation Checklist			
Pre-positioned Resources and Drive Away Kits			
Continuity Operations			
Re-establishment of communications Checklist			
Order of Succession Implementation Guideline			
Contingency Checklist for Vital Records and Databases			
Protection and Safeguarding of Vital Records and Databases			
Resource Acquisition Checklist			
Reconstitution			
Reconstitution Checklist			
Employee Advisories, Alerts and Instructions			
Green—Current and updated **Yellow**—Within three months due date to test, train, exercise, or review **Red**—Past due for test, train, exercise, or review			

Figure 12.

The final board will be the COOP Enhancement Suggestions Board, which is designed to allow all COOP program managers to provide feedback in terms of suggestions or ideas that will enhance any of the Job Aids in the Job Aids Board. This board will be interactive and will allow all program managers to share their ideas on suggested enhancements. The county COOP program manager or WebEOC controller will track suggested changes and support the updating process for all Job Aids in the county. There will be a drop-down window that will indicate the status of any enhancement or suggestion, and there will be a delivery date to indicate when an improvement is to be completed. This board will increase the information sharing among all of the 100+ COOP program managers in the county and will serve to help enhance all of the county organizations who have a COOP program.

The additional enhancements (legs) will complete the COOP stool and will greatly increase the overall COOP program effectiveness in the county. The current planning board already has had a major impact on our county's readiness, and the additional implementation and execution leg, along with the enhancement and suggestion leg, will only strengthen an already robust COOP program. The days of completing a COOP plan, putting it in a pretty three-ring binder, and then putting it on a shelf to gather dust are over. It is time to think of COOP as not just a plan but a program. Continuity operations works when a plan is implementable and executable!

Acknowledgement

Many thanks to the National Capital Region WebEOC Administrative Support Team comprised of Bruce Whitney, Joan Koss, and Ciprian (CHIP) Sufitchi, who took my simple spreadsheets and Word documents and made them into the wonderful WebEOC COOP board we use today. I also want to thank Gary Thomas, a veteran Montgomery County manager, who encouraged me to develop this COOP automated system tool. He is a man with great wisdom, and his encouragement was greatly appreciated throughout the development of this project.

ROBERT E. JONES, MBA, PMP, CEM, MEP. Over 34 years, Jones acquired experience and skill in project management, emergency management, international military affairs, and global business management. Currently, Jones is a program manager in the Office of Emergency Management and Homeland Security (OEMHS) in Gaithersburg, Maryland, where he works in the Emergency Operations Center, serves as the county Continuity of Operations Program Manager, develops and produces multiple emergency planning documents, and runs the HSEEP-compliant emergency exercise program. He is the Montgomery County WebEOC Administrator and Controller and serves on the National Capital Region WebEOC Policy Committee. Jones served 13 years as Executive Director of MCI's Global Project Management organization and 23 years as a Special Forces, Infantry and Civil Affairs officer in the U.S. Army.

Chapter 24: Strategic Procurement and Public Safety Communications: Managing Risk and Exploiting Opportunities
Public Safety IT and Currents of Change

BY ARTHUR S. KATZ

While IT may be a "support function," strategic planning in the public safety arena increasingly entails a significant IT component and IT dependency. First responders' ability to protect lives and property requires reliable, on-demand access to incident information and field personnel at potentially any time or location. That principle is, of course, a critical constant. But what will be re-shaping best practices for sourcing and implementing new IT systems and upgrades is the accelerating pace of change in the primary IT-related technologies serving the public safety function in an evolving commercial, social and political environment. Procurement for public safety agencies in this dynamic context for years to come will generate opportunities and challenges for executives that lead the IT support function.

The possibilities created by new technologies are shaping the ordinary capabilities and protocols for emergency services and law enforcement. Traditional land mobile radio (LMR) is likely to remain central to mission-critical voice communication for many years. IP-based systems, increases in available bandwidth, and the efficiency with which radio spectrum can be used are providing field personnel and dispatch centers with unprecedented types and quantities of data at increasing speeds. New software applications will continue to be developed that accelerate that trend. Next generation 911 services will integrate use of text, video, and email in responding to emergencies, and public safety use of social media will grow.

This reliance on more diverse and more complex technologies confronts strained government budgets at all levels of government. The challenge to do more with less is a further stimulus toward more efficient and more productive approaches to governance, design, and management of IT. Procurement planning for public safety IT must appreciate all of these trends, including by identifying secure and practical approaches to sourcing IT systems and services on a multi-jurisdictional basis or exploring joint ventures with the private sector for shared infrastructure or IT capacity.

Distinguishing public safety procurement

A sound procurement process can make the difference between realizing the rewards of new technologies and otherwise falling prey to unexpected costs, disruptive delay, and an impaired ability to pursue an agency's primary mission. Public safety agencies confront some special considerations when executing procurements for IT, for example:

- Optimal efficiency may dictate use by a jurisdiction's first responder agencies of the same communications networks and other IT, as well as by agencies in other jurisdictions. There are growing examples across the country of collaborative regional efforts to build and use voice and data networks. But in most areas of the country, there is little or no recent history of shared use on the scale that may be most appropriate in the future. New models for collective sourcing are and will be developed and refined as part of expanded multi-jurisdictional cooperation. Procurement exercises will have to operate within whatever administrative structures are developed to define regional needs and priorities. Increases in the efficiency with which the related IT is ultimately made available to affected agencies will justify multi-jurisdictional procurement; however, the process of RFP development and negotiation of definitive project documentation will be more complex, both as to developing technical requirements and as to authorization and management of dealings with vendors. This will be especially true when regional groups are gaining their initial experience with joint procurement.

- Most state and local budgets remain challenged by revenue shortfalls and growing obligations to provide public services. Funding for many IT initiatives and systems for public safety may be more vulnerable than some other proposed expenditures, because in battles over financial support for public safety services, reductions in IT line items may not draw public protest or political activism in the way that, for example, fire station closings or police officer layoffs often do. One reason is that unless a jurisdiction has witnessed significant tragedy as a result of technologically deficient communication or exchange of information, the impact on their lives of underfunded technology tends to be less perceptible to citizens than are cuts in the street presence of public safety front line personnel and facilities. The media is also less likely to focus on a delay in adopting improved IT capabilities than in the removal or reduction of a highly visible existing public safety service. For the average person, deferral of an IT upgrade has consequences that may not be measurable and that may not be easily understood (in contrast to almost universal community support for the well understood role of the first responders themselves). Additionally, major IT system changes like replacing a 20-year-old emergency communications network tend to have price tags that unless sufficiently subsidized by other sources (like federal grants) make them prime candidates for postponement.

- Federal grant money historically has been available in significant amounts for IT-related projects as part of disaster preparedness and maintenance of homeland security. In many procurements, securing such funds are a pre-condition for pursuit of a project, and eligibility requirements, therefore, must be complied with in the sourcing process.

- It is generally more expensive on a per unit or per user cost basis to pursue an IT-related project that must satisfy special public safety requirements for reliability, redundancy, interoperability, or regulatory compliance. Ordinary commercial-grade resources that may be adequate for other government operations or most private sector activities can support more extensive research and development (R&D), as well as mass production efficiencies. While some government operations in the military, intelligence, or finance areas face a similar need for advanced customization (1) in the case of finance, there are large private financial services and other companies that require similar state-of-the-art technology and whose business supports the availability of high-end commercial solutions, (2) intelligence is a much smaller market whose needs are also met by government-sponsored R&D, and (3) for the military, most purchases are developed and purchased by the federal government (or state or other national governments) with purchasing power that carries more leverage in vendor negotiations than the average public safety agency. Ineffective pricing strategies, therefore, can be a greater risk in public safety IT procurement, as can being penny wise and pound foolish in preparation of an RFP and for contract negotiations.

- The flip side of the preceding burden on public safety IT is that if more commercially available and off-the-shelf solutions can be developed that are adequate for public safety use, substantial savings could be realized. Intelligent sourcing strategies can seek out or instigate commercialized products and services that are or can be made adequate for public safety at lower costs. One example is the growing use of commercial wireless phone services to supplement public-safety grade handheld radios, especially when "smart phone" features offer functionality unavailable on the handhelds costing thousands of dollars more per unit. With the adoption of LTE as the standard for public safety broadband wireless communication and a greater focus on "modularization" of commercial hardware designed for selective special-needs upgrading, greater economies of scale will reduce public safety costs. Similar potential exists with private and public "cloud" developments that enable government users to reduce capital investment in IT infrastructure through increased reliance on sufficiently secure and reliable IT assets of commercial service providers, or government IT assets that may be shared across a larger number of jurisdictions or agencies.

- Current developments in interoperability and radio spectrum management (as well as budget considerations and economies of scale unrelated to more use of commercialized technology) are very likely to push public safety agencies toward both greater multi-jurisdictional (including state- and regionally managed) procurement. While this may also be a trend outside of public safety (or IT), in the case of public safety, there are particular technology-based and (primarily federal) policy initiatives providing the push toward collaboration and consolidation. A fundamental trait of the 700 MHz public safety shared wireless broadband network (SWBN) authorized by recent federal legislation is that, by design, it will assure interoperability across the entire nation. That increases the logic of cooperation across jurisdictions jointly to build out regional portions of that network and facilitate additional regional efforts toward efficient use of other existing networks as they are integrated and updated. This will impact how related IT procurement is conducted, both for purposes of securing SWBN infrastructure and for the integration of that functionality with other state and local systems and applications.

- Public safety communications is ripe at the moment for a substantial increase in the role that "public-private partnerships" (PPP) play in the roll-out of next generation communications and related IT projects. The permutations for mutually beneficial PPP collaboration include public sector use of commercial network infrastructure, private carrier services supplied to the public using government-owned infrastructure, joint public ownership, public sector infrastructure partially funded by private capital in exchange for network access, and others. The procurement issues arising from public/private transactions affecting public safety functions are distinct from and in addition to those related to joint use and support of a network or other IT project by multiple government jurisdictions. Among them is the likely need to address a scope of both profit-motivated and operational concerns that would not arise in a conventional system purchase or vendor services contract.

- Opportunities for public sector capital and operating cost savings arise in the sharing of emergency communications networks with "non-critical" government services (e.g., wireless water meter monitoring), where data transmission over wireless broadband may replace more labor-intensive methods of information gathering. Jurisdictions that have pioneered such programs are reaping the benefits of greater utilization of the capacity of networks with minimal concern that public safety priorities will be compromised. The low marginal cost of the lower priority usage generates savings for the jurisdiction that effectively reduce a jurisdiction's net cost for a new data network. Although it is very possible to add such non-critical usage long after a network is completed, contemplation of such usage may well be important as CIOs assess the projected return on investment of public safety IT investments.

- Similarly, quasi-public entities such as utilities have needs (e.g., electricity grid monitoring and control) that will support user fees for access to government networks, an in-kind contribution of infrastructure, or both. Federal rules are being liberalized in these areas, but priority and pre-emption practices will require negotiation and will be facilitated by further technological developments. Uncertainties in the procurement process would be reduced by advance or parallel-track agreements on asset ownership, development, and management, usage-based fees or other compensation, and similar topics that arise with the particular plan for joint operation or use of limited, critical resources.

- The current trend of "moving to the cloud" should be expected to attract public safety agencies no less than other public sector players. However, as with the growing capability of first responders to access data in the field through wireless connectivity, public safety must address for cloud-based applications their unique security concerns regarding storage and transmission of information related to law enforcement and anti-terrorism, personal health data used by emergency medical services, and other potentially sensitive information such as non-public blueprints or other incident-relevant data that might be transmitted to fire and rescue personnel.

- Collaboration among jurisdictions or in PPP arrangements may increase points of vulnerability to the introduction of viruses, hacking, and terrorist attacks, even as a trend toward shared systems may reduce available redundant or alternative IT resources. Procurements will not only need to identify leading-edge capabilities when crafting specifications and evaluating system design, but assure that as new security threats emerge or legal requirements change, prompt action will be undertaken to maintain IT system integrity.

What is a successful procurement?

How does one plan an effective procurement strategy in a public safety IT environment facing fundamental changes in principal technologies, infrastructure ownership and management, joint venturing among the builders and the users of IT systems, as well as an evolving regulatory framework and legislated mandates on infrastructure development and administration? The following are some key indicia of a procurement exercise's having followed success-oriented practices, and notes on some timely practices:

1. At execution of the related purchase order/agreement, the agency customer understands what it is committing to buy.

    - Some might consider execution of the agreement to be a bit late in the game as the deadline for understanding the intended purchase. That's true in the sense that the procurement process and related RFP should (including discussed above) from the start reflect attempts to obtain and use relevant technical, economic, business, and legal information to intelligently shape purchase priorities, as well as to identify and establish the administrative structures appropriate to manage the procurement.

    - Most RFPs for complex IT purchases are and should be requests for functionality and performance, not specific product models or software releases (although there may be compatibility parameters referencing product models and releases). It is the review and evaluation of bidder responses and the subsequent contract negotiation process that brings the project into much greater definition, especially at a time of rapid change. Even if a purchase is made subject to a later final design, the public safety customer should understand what the contract certainly will supply and what remains subject to variation so long as agreed performance criteria are ultimately satisfied. In a situation where PPP alternatives are being considered at the same time as fundamental technical parameters, the procurement process may need to allow for forward progress on one front before being able to move on another, and applicable purchasing rules will need to have anticipated that.

- In the case of projects involving significant customization or introducing innovative administrative or ownership structures, the procurement exercise will be associated with a learning curve that educates both customer (no matter how many jurisdictions comprise the "customer") and vendor (including candidates for co-ownership or management of IT assets in a PPP context). That education will refine what the interested parties view as technologically feasible, economically viable, practically manageable, and otherwise adequate to meet their respective goals for the purchase transaction. In this context, agencies should be careful to reserve for themselves a substantial flexibility to revise proposed technical and other terms and conditions as the procurement moves forward. The more groundbreaking the procurement seeks to be, the more flexibility needs to be retained as initial RFP documentation is issued and vendor discussions commence. It may be more appropriate to issue a more exploratory request for information rather than request for formal proposal. This permits the eventual RFP to reflect a greater awareness of vendor perspectives, while also allowing progress to be made (if needed) to work out joint venture/power sharing arrangements among cooperating jurisdictions, continuing any needs assessments for prospective project users, analyzing or pursuing financing sources, etc.

- The goal of the procurement should be to sign project documentation that generates the specified system on budget and on time. Until the contract is inked, to the extent procurement rules permit, the purchasing exercise should be flexible enough to allow the agency to act based upon its growing knowledge. What it learns prior to execution can be used to address material considerations that may not have been appreciated at the time of RFP issuance or even at vendor selection. Vendors should be advised that the agency's goal is to continue to refine the procurement as it gains knowledge, both to avoid vendor protest based upon contrary expectations and hopefully to secure vendors' active cooperation in educating the agency in good faith before any contractual commitment is made.

2. The agency understands the vendors' RFP responses and positions on terms and conditions, including as incorporated in the definitive purchase contract.

    - This point may seem so obvious as to not be worth mentioning, but if it were emphasized more, there would be fewer disputes and change orders arising during the life of purchase contracts.

- Those who negotiate purchase contracts don't always share the same philosophy as to the scope of contractual terms and conditions or the scope or detail of the contingencies that should be explicitly covered. But most would agree that in a procurement that may be breaking new ground and, therefore, confronting a lack of testing thinking or market convention, it is prudent as a purchaser to brainstorm, identify, and assess the likelihood and severity of the risks arising in the atypical aspects of the transaction. An agency should expect that the commercial party across the table has done that and proposed terms that address such risks as they impact that vendor (or PPP private sector "partner").

- Vendors should not be expected to deliberately mislead or confuse prospective purchasers (although, unfortunately, it has been known to happen). If requested, almost all vendors will do their part to clarify language and to elaborate on potentially relevant details to reduce ambiguity and misunderstanding. But vendor teams are often supplied with "boilerplate" or provisions borrowed or poorly adapted from past sales efforts. This is a greater risk and particularly problematic in the case of innovative procurements. Moreover, those team members don't always focus on vendor-created terms that might present risk to the purchasing agency but not to the vendor. Moreover, persons on both sides of the transaction may hesitate to question information that doesn't seem quite right, out of fear of appearing ignorant, or because they assume someone else will make the point or raise the question if it is valid and worth attention. Public safety IT transactions are hardly unique in that risk, but it's worth keeping in mind that the varied constituencies that are gathered even for a major public safety-oriented procurement may include several representatives that have not participated in a similar process before. In managing the evaluation process and working with the negotiating team, a project manager should regularly encourage questions or expressions of doubt or confusion. Most veterans of larger procurements can cite examples of "ah-ha moments" where an emperor of a vendor proposal was determined to be seriously underdressed after an inexperienced evaluation team member raised a point others had overlooked. While there may be times when a procuring entity is served by a silent or ambiguous contractual provision, that should be a conscious determination and not a lucky accident.

- Even if vendors don't seek to mislead, they have their own set of goals, priorities, and resource constraints. Bids and vendor contract forms may include "playbook" language or terms attributable to "policy" that have been crafted to protect the vendor in ways that may not be obvious, even to some members of the vendor team. Don't count on the vendors to highlight these items, but do expect vendor negotiators to be able to explain them or to supply justification from headquarters so that the purchasing agency may take an informed position on the subject matter. The less routine the transaction, the more likely the vendor playbook gospel has failed to capture nuance, and further editing might well serve both parties.

- Notwithstanding the above comments that anticipate vendor-drafted forms and other terms, procuring agencies should strongly consider developing their own purchase contracts. Unfortunately, most purchasing department standard terms, even for IT transactions, may be inappropriate to govern multi-jurisdictional, PPP, or early adopter-type IT solicitations. Depending upon the size of the anticipated transaction, agencies should consider whether to reject most vendor-developed documentation in favor of an integrated, customized contract. While time will be required for the drafting of such a document, it also will serve as a catalyst for internal research and polling of relevant constituencies. It also will turn the contract document into a basis for competition among vendors so far as accommodation of preferred agency positions and avoid the time-consuming need to try to "compare apples to oranges" as different vendors submit their own proprietary contract forms.

3. Bids are evaluated effectively and fairly in order to appreciate the material differences among vendor proposals and variances in the bids from RFP requirements (and other permissible considerations that are identified during the procurement exercise).

- The methodologies for evaluation and scoring of RFP responses vary widely among purchasers and professional advisors, sometimes dictated by purchasing rules or precedent. Many include highly subjective weighting and scoring that may not be unreasonable, but also can't be called objectively "correct." Ideally, such scoring does not so much control a vendor award as guide the internal discussions on the (dis)advantages of one vendor proposal relative to the competition. Do not underestimate the benefit of having the evaluation team that represents important first responder and other stakeholder groups (including IT professionals) present their numerical scoring to the group along with verbal or written justification. This can expose inconsistencies in the application of the scoring criteria that contaminates results. It also draws the attention of the entire group to considerations that may have been overlooked by some, and may shift opinions or give some team members a better ability to apply their special (technical, front line, administrative) perspective to the evaluation.

- More important than framework of the scoring system may be the scope of the items explicitly considered in the scoring. Ranking vendors using categories like "vendor experience" without incorporating further guidance on relevant sub-elements (such as number of similar systems implemented, experience and tenure of the proposed implementation team, size and similarity to the soliciting agency of other customers, etc.) may yield meaningless scores because the members of the evaluation team apply the criteria differently. Because of the risk of an eventual protest by losing bidders, it is important that evaluation guidelines, like the use of scoring, appear rational, well understood, consistently applied, and not given more weight than can be defended as logical.

- The preceding point is especially relevant in light of the varied constituencies that may wish to be part of the evaluation of a major project such as an emergency communications system (as discussed below) or a novel procurement involving a PPP arrangement. Proper attention to constituencies can be essential not only to success in specifying project scope, design, and performance, but also to the political success of the procurement and subsequent implementation. For example, each constituency may have a different concept of what elements of "vendor experience" are most critical to emphasize. The evaluation process should be designed to flush out those special perspectives so that they may be given proper consideration. Ideally, the process is transparent enough to enable education and consensus-building, yet sufficiently private to avoid needless embarrassment or alienation of constituencies whose views do not prevail or are perceived as uninformed or too self-interested.

4. Vendor commitments provide reasonable confidence that a vendor will timely perform its material obligations, including across important contingencies.

- Agencies should not be shy in asking their internal experts or outside advisors to compile and include in an RFP performance specifications that are appropriate to assure that mission critical services will be fully supported under emergency conditions. Ideally, there will be awareness on the project team of appropriately aggressive parameters for similar systems, including those recently purchased by other jurisdictions or the subject of another current procurement.

- Although a role of a purchasing agency's outside advisors is to contribute awareness of the technologies and performance commitments that other jurisdictions are securing from vendors, too often internal agency staff and management fail to exploit the wealth of information available to them for the asking from their counterparts across state, county, and municipal lines. Vendors usually will be happy to identify to whom they have recently sold similar systems and supply customer contact information. Subject to confidentiality obligations, most of the time CIOs and project managers will be happy to discuss with their counterparts their procurement experiences.

- It may or may not be helpful to identify another jurisdiction to a vendor when seeking a particular performance promise. Vendors generally will admit if the sought commitment has in the past been given to another customer, but they may try to distinguish the other customer as receiving the benefit as a result of having a larger project, broader relationship, or strategic position in the vendor's target market. The vendor may even attribute previous agreement to the provision to a "mistake" by the vendor that, therefore, won't be repeated. Start the conversation and negotiate from there.

- Don't be afraid to ask for something that the agency needs or that it feels is reasonable in the circumstances, even if the vendor may never have agreed to the particular term. Worst case, the vendor's response is negative, or success may require "horse trading." Changes in technology regularly make feasible and reasonable performance levels that were previously unachievable, and vendors should be open to tracking such evolution.

- Although it occurs fairly infrequently, situations do arise where a contract is awarded and then a vendor fails to perform or is terminated for other reasons. This possibility should be anticipated and addressed in every contract, with the related risks assigned to one party or the other. Additionally, because it is the first responders' job to perform under adverse and extreme conditions, attention should be paid to the management of transitions from one vendor to another or from one system to another so as to minimize any compromising of mission critical capabilities.

5. Allow for mid-course corrections and improvements.

    - Even a thorough RFP and extensive negotiation of purchase terms and conditions will not include a script that addresses the timing and nature of all future events that may ultimately affect the project. But terms and conditions can be devised that at least establish a framework and procedures for addressing non-routine events.

    - A pre-agreed escalation procedure can be helpful in resolving disputes, because it generally sets a mandatory schedule that incents quick resolution of disagreements before superiors may blame subordinates for creating or failing to diffuse the problem issues. Additionally, when higher-level executives do become involved, it is often possible both to find solutions within the broader business relationship between the parties, or alternatively to conclude that the problem be handed to litigation or other counsel. Either course enables regular project personnel to maintain their focus on regular project activities.

    - Clear change order procedures also can save time and avoid disputes, first by indicating agreed circumstances where a vendor is clearly entitled to a cost-increasing change and how those costs are to be established. In other cases where a vendor may seek a change order but the purchasing agency maintains that some or all of the change costs should be borne by the vendor, a change order process still can facilitate prompt identification of reasonable costs and planning to execute the change, allowing the allocation of costs between parties to be handled on a separate track, in dispute escalation, by financial function personnel, and/or at a later time with unrelated other economic issues.

    - If a project system anticipates updates or upgrades in line with historical experience or anticipated later-generation technology, there also may be an opportunity to pre-negotiate or "collar" (set a minimum and maximum for) some costs for planning and budgeting purposes. An agency needs to be careful not to agree up front to costs that might have been lower had it waited until the time of the foreseeable upgrade/update, but the agency's negotiating leverage may be greatest at the time of the original system purchase.

6. Identify, solicit, educate, and involve important stakeholders.

    - The public safety arena presents a complicated and demanding mix of interests and cross-currents that should not be ignored in a procurement exercise if it is to proceed smoothly. Most relevant constituencies fall into one of four categories: operations personnel, executive function personnel, political actors, and external regulators. The procurement process must accommodate each, but the timing and approach differ significantly. To plan a procurement process that does not pay attention to important constituencies from RFP design through vendor selection and a definitive contract invites not only late-stage delay, but can force a reopening of settled contract issues or even a re-solicitation of bids. Rarely do such scenarios generate kudos for the procurement team, because they are largely avoidable with good planning.

- The "first responder" constituency of police, fire, EMS, and other services perform different functions in different environments and cannot be assumed to speak with a single view on what the highest priority features of an IT or communications system should include. Personnel in the field might also differ from those in dispatch centers or in senior agency management as to what are essential features.

- Jurisdiction executives with C-level authority and responsibility, as well as the jurisdiction's CIO or equivalent, will have yet another perspective. To the extent that there are potential opportunities for more than one jurisdiction to collaborate on a project or to explore a project development with a private sector partner, senior jurisdiction executives should be consulted at an early stage and regularly thereafter. A large or politically sensitive project may justify appointing a senior management "steering committee," in addition to a project team and evaluation team, as a means of building consensus between first responder agencies and other departments on project priorities.

- While voters and elected officials may only focus on IT-related public safety issues rarely if at all, their approval may be required for project appropriations, project bond issuances, or general budget matters that affect project feasibility. Depending on the project, it may be congressional rather than state legislature or local council action that is most critical to funding or authorization. During the procurement process, the project manager and CIO should be prepared to provide clarification required by elected supervisory bodies or to address public misconceptions regarding the project. It may be best if a single public spokesperson is selected to deal with media inquiries, keeping in mind that this may not be a skill that was considered critical in the project manager selection process.

- Finally, federal and state laws make regulatory authorities ranging from the FCC to the Interior Department other potential constituencies, depending on the nature and scope of a project and the location of its infrastructure. To the extent permits, licenses, or special waivers of applicable rules are required, a project's schedule and administrative costs may be affected. For that reason, these requirements should be identified and reviewed with vendors so that responsibilities can be allocated and costs properly estimated as part of the procurement process.

7. Field the right team.

- There is no single formula for building an effective team of internal IT staff and outside advisors. Larger jurisdictions tend to have deeper internal resources and be less in need of outside professionals, particularly for smaller projects or upgrades to existing systems. Smaller jurisdictions may lack internal expertise, but also pursue projects that have budgets large enough to support a diversity of outside professionals.

- The most critical role likely will be the agency's project manager. There should be as much continuity in the project manager as is practical through procurement and implementation, so that the "institutional memory" of the RFP and contracting exercise is not lost as practical implementation begins. A project manager who can advocate for a jurisdiction based on a continuous awareness of the evolution of specifications and other contract terms can generate benefits from the contract from between its lines. He or she is also in a position efficiently to consult with others that played lesser or more fleeting roles at the procurement stage.

- The technical advisors may be internal staff, but for larger projects it is generally better to supplement internal staff with IT, engineering, or other consultants who have experience with procurement of the most current relevant technology. Sometimes, the consultants restrict themselves to purely technical roles, advising on specifications, system design, and operational questions. Other consultants will offer their services to handle both technical and "business" matters, which may include general terms and conditions, and may include financial terms if the finance or purchasing department of the jurisdiction is not equipped to handle those aspects of the procurement.

- Legal advice is sometimes handled by internal counsel if similar projects have been handled in the past, and the demands of drafting and negotiation are not incompatible with internal lawyers' other daily responsibilities. Regular outside counsel on legacy systems may have the required skills for consultation on matters like spectrum licensing, environmental issues, and intellectual property matters, but not broader transactional experience. For major system acquisitions and less orthodox transactions, a jurisdiction is likely to want counsel that is familiar with transactions with similar systems or other highly customized IT and communications products.

- There may be some overlap in roles and expertise, for example along the spectrum of contract lawyers working on general terms and conditions, business consultants that have collaborated in development of such terms, and engineering personnel drafting technical attachments to the contract that ultimately must be integrated with the general terms. All of the above and internal staff may be present at the negotiating table with vendors. The project manager should clarify to the negotiating team who will be the primary voice and master of ceremonies opposite the vendor team.

- Multiple outside advisors can run up large administrative expenses, and the number of advisors and extent of their roles should be weighed at the beginning of the procurement process and prospective budgets compared to estimate project cost. Properly managed as part of a plan, advisors can be used efficiently in their areas of greatest value, with the outlay being the equivalent of an insurance premium paid against the risk of technical and contractual deficiencies that might deny the purchasing agency principal operating benefits of the project. Advisors should offer their views as to their own highest and best use in a procurement and cooperate on designing a role for themselves in the exercise that will make optimal, cost-effective use of their particular skills.

Innovative IT procurement for financial and operational benefit

This chapter has been less of a comprehensive primer in procurement tactics than a briefing on special considerations when approaching sourcing events for IT serving public safety agencies. In the same vein, it's worth considering the potential that CIOs and their staff will have to use procurements to harness some of the forces that are shaping what will be the largest public safety IT procurements.

In the next decade, there likely will be a quantum leap in the scope and sophistication of IT system governance as it impacts public safety. Consider the upcoming 700 MHz public safety interoperable broadband network.

- The SWBN will develop billions of dollars of IT infrastructure across the country. The resulting network will be overseen by the national FirstNet authority, which will have the benefit of a formal representative public safety advisory board (as well as public safety experts serving on the FirstNet board of directors). Design of the SWBN begins with nationwide state-level plans. But the network and its use will be integrated with existing and to-be-developed local communications and IT assets.

- Private ownership by carriers, utilities, and others of some assets essential to the SWBN likely will become commonplace. The authorizing legislation encourages public-private arrangements in order to facilitate the build-out and to conserve resources. While there already exist scattered examples of shared public-private infrastructure or public-private use of either public or private assets, these remain novel concepts for the great majority of jurisdictions. The economics, operational procedures, and security considerations will present issues of first impression to most public safety IT decision makers, and a manner of "doing business" with the private sector will have to develop that requires broader skills and experience in the public safety IT procurement process than for the execution of a fees for services contract (or even the building of a traditional local government-owned emergency communications system).

- While the authorizing legislation anticipates that much of the national framework for addressing these public-private complexities will be modeled at the federal level with FirstNet, there will be no escaping the need to manage around local issues. If the enormous potential of public-private collaboration is to be explored and exploited, at a minimum public safety agencies (or the coordinating IT function) will need to engage on the issues. Hopefully, there will be local level initiative and involvement in the identification and pursuit of public-private partnerships.

- The fact that the SWBN is a national project with FirstNet oversight and deliberate public safety input will greatly enhance the ability of localities to draw upon other relevant experience in addressing local partnering issues. Moreover, the SWBN legislation focuses on supporting research and development and other resources that will enhance the ability of a region to benefit from the SWBN activities undertaken elsewhere. But to conduct effective related procurements, states and localities still will need to plan and advocate so as to reflect their respective actual circumstances, including political, fiscal, and technological realities that may not have precedent elsewhere or that offer the opportunity to improve upon past solutions.

The SWBN offers an example of legislated change that will have an essentially universal impact on public safety IT planning for all jurisdictions. It's useful for illustrative purposes for that reason. The many administrative challenges and possibilities it presents will appear in more generic form in unrelated procurements: sharing services or infrastructure with private sector entities to reduce capital or operating costs, coordinating across jurisdictions for the sake of the benefits of technological compatibility and economies of scale, and establishing governance for IT-related systems across wider and more complex user and ownership profiles than in the past.

Procurement in this context requires more preparation, a consideration of new external forces and factors, and a need for multi-dimensional orchestration of the events that become important to contract award and implementation.

- Most major procurements do not solicit vendor interest with an RFP until viable funding options have been identified. Historically, for public safety that has frequently meant securing an award of federal grant money and required matching funds. Beyond that, the growth of "partnered" projects will, on the government side, require identification of the jurisdictions that will share ownership of project assets and (if a different group) those that will share operational control and management authority. Beyond a need to identify the potential geography of a project, at least preliminary agreement will be needed on the governance rules, because the ability to support the governance structure may create additional RFP specifications for system or vendor performance.

- Partnering with private sector partners will impact procurement in other ways. If the private party is supplying some or all of the capital, the agency will require correspondingly less funding (but may want to plan for the possibility that the private entity will withdraw, especially if grant requirements impose deadlines that risk loss of grant money if the procurement does not proceed without interruption despite loss of private money). But to the extent the private interest(s) will be exposed to the infrastructure or operating costs of the project or have vendor requirements that are not covered by agency standards, their positions will affect both RFP content and vendor negotiations. Even where the partnering arrangement does not involve joint ownership or management of a network (for example, sale or exchange of excess capacity on the government or private sector system), there will be issues of compatibility and planning for adequate capacity that will need to be identified and quite possibly addressed in the RFP or later in the procurement in order to attempt to optimize the cost-effectiveness for the agency of the public-private arrangement.

- At the CIO level, the strategic and management implications that are reflected in procurement documentation would be fundamentally altered by the choice to "partner" on an IT system under alternative structures. In contrast to traditional options of either constructing a government-owned and operated system or embracing only a contract for vendor-provided services, there are hybrid models such as having agency-managed services hosted on non-government infrastructure, retaining a carrier or other vendor to managed services delivered using government-owned infrastructure, integration of public and private infrastructure for provision of services to the government and/or special classes of users or the public at large.

The alternatives present difficult tradeoffs among matters related to cost, access to capital, security, priority access, allocations of capacity, control, potential liability, and the availability to public safety of "commoditized" yet adequate hardware and applications. But the potential upside of intelligent and creative approaches makes the challenge of exploring the alternatives compelling. The unique requirements of first responders may eliminate some options that would work for organizations with less demanding requirements, but it is the criticality of public safety services that also justifies the effort to try to find solutions that can extend the benefit of emergency communications coverage and IP-based and other IT capabilities to all corners of the country.

Those guiding IT procurement for public safety find themselves living in very interesting times and with the chance to make far-reaching contributions.

ARTHUR S. KATZ is the principal of Katz Law Office, P.C. (KLO), a law firm focused on the procurement of communications networks and other IT resources by public sector entities and Fortune 500 corporate consumers. KLO serves as primary or special counsel providing strategic, negotiation and documentation support, directly and as a subcontractor to other professional advisors. Mr. Katz spent almost 25 years in internationally recognized law firms, where he advised clients on diverse commercial, technology, and financing transactions. He was primary draftsman and a lead negotiator of the contract documentation that launched New York City's trailblazing NYCWiN wireless emergency communications network. Since KLO was formed in 2007, he has worked with state and municipal government clients across the country on purchases and upgrades of public safety communications systems.

Chapter 25: The Story of Technology, Integration, and Collaboration to Support Fire Incident Response and Command

BY DAN RAINEY AND JASON MCKINLEY

Overview

Fire responders have long relied on outdated paper maps and hand-drawn pre-plans while responding to emergencies, leaving access to powerful Geographic Information Systems (GIS) back at headquarters. Fire personnel in the field did not have access to the wealth of GIS information currently available from the city, other agencies, and commercial data that would enhance their ability to effectively respond to an emergency. Further, command is stretched to track and deploy resources across a large geographic area. Computer-Aided Dispatch (CAD), Mobile Data Computers (MDCs), and Report Management Systems (RMS) were standalone, silo systems with no interfaces. Up-to-date information was limited on scene to radio communication, and data was re-entered by hand into the RMS after an emergency, leading to duplicate efforts and data quality issues.

The city of Ann Arbor, along with its partners—the Huron River Ambulance Authority (www.HVA.org), OptiMetrics, Inc. (www.adashi.org), Xerox (www.firehousesoftware.com), and ESRI (www.ESRI.com)—assembled and deployed a comprehensive software platform for managing, communicating, responding to, and reporting critical incidents, providing a common operating picture and improving situational awareness. This integrated public safety platform has evolved from a solution for a single department to one that potentially can be utilized regionally. Currently, plans are under development to deploy this solution to various agencies across Washtenaw County, hosted by the City of Ann Arbor. This is a real success story of technology, integration, and collaboration that can only enhance response to emergency calls for services.

The problem

The dispatching of emergency services had long been handled by numerous stand-alone dispatching centers across Washtenaw County, for which Ann Arbor is the county seat, without the aid of a common operating picture, resulting in a lack of situational awareness and, at times, a less-than-optimal deployment of available resources. In 2002, the City of Ann Arbor combined its fire and police dispatch centers into a single city dispatch, and over the next several years, the fire dispatcher positions were phased out. However, optimal fire dispatch requires dispatchers to be trained to be able to quickly analyze the nature of a fire emergency and ensure that the minimum quantities of required apparatus are engaged. Without proper training, dispatchers, erring on the safe side, often sent more equipment than necessary to a scene.

In 2010, in an effort to more effectively manage the use of fire resources, the city opted to partner with a regional fire dispatch and ambulance consortium, the Huron Valley Ambulance Authority (www.HVA.org), for fire dispatch. Currently, HVA provides fire dispatch and 9-1-1 paramedic ambulance service to about 1 million residents in all or part of eight counties in southeast and south central Michigan, including fire dispatch services for 11 fire departments. Having fire and EMS dispatching within a single, larger consortium has saved the City of Ann Arbor response time, improves shared information, conserves limited resources, and reduces risk. The HVA provides fire dispatching at cost to the City of Ann Arbor, and pricing has only risen from $13.02 per alarm to an estimated $15.62 per alarm, with fewer alarms being dispatched because of the use of trained fire dispatchers.

As is still the case in many jurisdictions across the country, City of Ann Arbor fire responders have long relied on outdated paper maps and hand-drawn pre-plans for routing directions while responding to emergencies. Even with the proliferation of technology during recent years, little had changed in the way Ann Arbor firefighters found their way to a fire or EMS call for service. The city had invested in key fire technology over the years, including computer aided dispatch services, incident response, and incident command systems, and a fire record management platform, but these systems were still standalone, silo systems requiring manual processes—some on scene and some after scene—to complete the necessary information integration required to support the business processes required of the fire department.

Solution

To provide the fire department with proven incident response and incident command capabilities in a networked emergency management system for coordinated use across all agencies, departments, and staff levels, the city selected the ADASHI incident response and command suite of ADASHI Dispatch, ADASHI First Response, and ADASHI Command Post (www.adashi.org) software. The department is currently live with the suite on 10 of the city's fire vehicles across five stations. This incident response and command solution is sophisticated, yet easy-to-use, decision-aid software specifically designed to meet the needs of the city's emergency responders by tracking and displaying the response unit's current position, creating a turn-by-turn route to the scene, providing all-hazards incident information management solution for first responders, and adding mission-critical tracking, management, communication, logging, and reporting features. By combining the ADASHI software with the power of ESRI's state-of-the-art ArcGIS software and utilizing both City of Ann Arbor and Washtenaw County GIS resources, a wealth of local information is placed at the fingertips of emergency responders, including the locations of fire stations, hydrants, schools, and hospitals, as well as the ability to identify potential hazard sources, infrastructure, evacuation routes, and more.

When the City of Ann Arbor moved to the regional fire dispatch services provided by HVA, the Fire Department also began using the HVA's customized Computer Aided Dispatch (CAD) as the department's fire CAD. The city's Information Technology Department facilitated the development of two interfaces from the HVA's Fire CAD system. One interface runs to the fire department's records system, Xerox's FireHouse. When a dispatcher closes an incident in CAD, data is transferred into the records system using the FireHouse CAD Monitor application. The time-consuming manual task of inputting detailed incident data, such as arrival times by apparatus, is now automated, allowing a firefighter to focus on the narrative of the report. The second interface is from HVA's Fire CAD system to ADASHI's CAD interface, providing address and routing information to the ADASHI suite, a live CAD feed with updated scene status, and coordinates to the scene. In a third interface under development through a collaboration project between Xerox and Optimetrics, all FireHouse data, such as pre-plans and permit information, is synchronized back to the mobile computers. Critical scene information—such as pre-plans, hazardous materials, and building layouts previously trapped back at the fire department's headquarters in FireHouse—will be readily available on scene during an emergency.

The diagram below depicts this flow of information between all the systems:

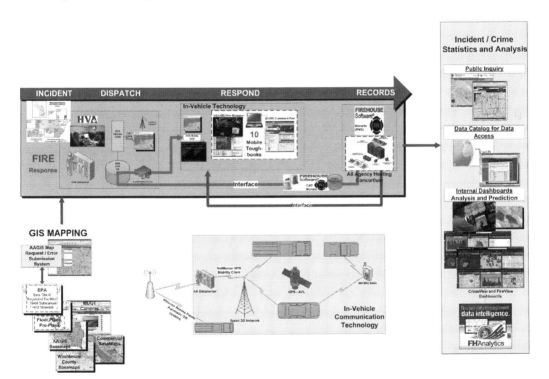

Figure 1.

By leveraging the relationship with HVA and HVA's customized fire CAD platform, the City of Ann Arbor Fire Department, working in partnership with the city's Information Technology Department, has successfully deployed a comprehensive software platform for managing, communicating, responding to, and reporting critical incidents. This best-of-breed solution is based on the real-time integration of HVA's customized computer-aided dispatching platform, emergency incident response and command software from ADASHI, records management from Xerox's FireHouse RMS, live mapping services from Geographic Information Systems leaders, ESRI (www.ESRI.com), and locally produced GIS datasets. This integrated system is live in Ann Arbor, where the city's firefighters have traded in their trusty paper maps for a reliable modern mobile computer connected to the Internet.

After the success with the City of Ann Arbor, the neighboring City of Ypsilanti expressed interest and joined on the same system. Other fire departments from Washtenaw County now have taken interest, and plans are being developed to deploy this solution countywide, hosted by the City of Ann Arbor. This is a real success story of not just technology and integration, but also collaboration and trust that can only enhance local government's first response to emergency calls for services.

Barriers

Both human and technical barriers are common with the implementation of any new system, and should be recognized and overcome early in the project. This is particularly critical with incident response systems in that after "going-live," these tools are being used in life-or-death situations. Any technical hiccups (barrier) in the system will have a negative effect on user acceptance, thereby creating new human barriers. How effectively you address these issues and communicate progress is directly correlated to your level of implementation success.

Barriers recognized early in this project were:

1. Incident location: The incident location is determined by the CAD system and a dispatcher speaking with a caller. The latitude/longitude of the location is then transmitted to the mobile units, and a best route is calculated. We found that over 90% of the time, the location is spot-on; however, the location can be very incorrect when misplaced. Complex campus locations or disoriented cellular callers may present significant routing challenges.

    - Solution: Rather than blaming the system for incorrect incident locations, training was required to demonstrate how the system can be used as a tool to zero into the actual incident location using various GIS mapping layers on-the-fly, while approaching the incident location.

2. Human nature resists change: To the extent possible, do not underestimate the reliance on paper maps to locate an incident. Technical issues were seen as a reason not to

change, and some staff would literally close the computer at the beginning of shift and revert back to the large binders of hand-drawn paper maps.

- Solution: Show top management support, react to technical problems quickly, stress the importance of their concerns and input, and highlight the proper behaviors.

3. Global Positioning System: When the GPS signal is lost, usually when parked inside a station, the system loses the ability to track a vehicle until satellite signals are regained, typically a couple city blocks after leaving the station. Further, the vehicle displays a location-unknown icon to command while parked at the station.

- Solution: Install external GPS antennas on each station roof that will amplify the GPS signal in the station. This solution allows vehicles to maintain GPS connectivity and broadcast a correct location while parked in the station.

4. Cellular Signal Lost: In a few instances, cellular connections are not available within a station or in some dead zones across the city. Often, it was reported that the software was not functioning when it was really network connectivity issues.

- Solution: External antennas were installed on all trucks that help amplify the signal received and are a large improvement over the internal network cards in the MDCs.

"Backdraft"

With new technology adoption, often we are given one shot at success. Projects either implode or explode, depending on a variety of factors. A backdraft is an explosive event at a fire resulting from rapid re-introduction of oxygen to combustion in an oxygen-starved environment where combustion has slowed, for example, opening a door to an enclosed space. This is a good analogy for this project where early project barriers had slowed the project, and it was at risk of failure. Change is difficult, particularly when the technology does not perform exactly as promoted right out of the gates. Quite frankly, firefighters can quickly lose faith in the project, armed with the technical difficulties as a way out of using the new system.

As the technical barriers were resolved, the system started gaining success. Rumors spread that the system actually assisted in responding to difficult incident locations, such as university campus buildings and rural HAZMAT calls outside of the city limits. As success stories began to outweigh technical glitches, a "backdraft" occurred, and project momentum accelerated. Satisfaction with the system shot upward, and the tool was adopted as a part of standard operating procedures. This backdraft continues on a regional level as other local departments learned of the success of the system. The project was presented to the Countywide Fire Chief Committee hosted by HVA, which is now approaching the project on a regional basis.

How it's working

When a call for service is received by the HVA, the location coordinates are immediately sent to the ADASHI software running on an MDC in a fire truck or in a battalion chief's vehicle. Detailed CAD information is streamed to the MDC, and the quickest route to the incident is calculated and displayed. Real-time GPS tracking services allow all units in the network to visually see each other on the ADASHI/ESRI local map interface. This is especially important during a mutual aid call with another jurisdiction, aiding the partner agencies in locating the scene. Detailed GIS information is readily available via touch screen, particularly valuable in both dense downtown locations and in locating rural addresses.

Today, the HVA hosts a central ADASHI server within its datacenter located with the central dispatch. This configuration allows the HVA to monitor all participants' vehicle locations on a single large monitor. It also allows them to monitor active incidents and coordinate any situations with multiple active incidents, ensuring the optimal use of available apparatus. New participants are easily added to the system by a simple change in configuration.

Figure 2 and Figure 3 depict the system in action on a touch-screen MDC:

Figure 2. Overview map of active calls for service and all available units by their status.

Interfaces are in place between the HVA CAD and ADASHI system, installed on an MDC, as well as the HVA CAD and the FireHouse RMS, located centrally at each fire department. Up-to-date GIS data flows into the mobile units, enhancing the past process of manually updating paper map books. The interfaces run on top of a series of network pro-

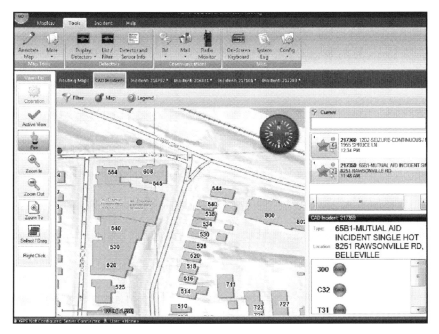

Figure 3. Detailed incident data, response units by status and GIS data readily available.

viders and wireless access points. The general process flow is that an incident comes in to dispatch, and a new record is created in the HVA's CAD system. The incident information is then "pushed" to all MDCs running the ADASHI software. When a fire station is notified via VHF or IP-based notification system of a new call, they open the MDC, and the incident has automatically been assigned to the correct apparatus and a suggested route has been calculated to the scene. During the incident, all CAD updates are broadcast to the MDCs, and the firefighters can instant message back to dispatch or to any other active vehicle via the touch screen. After the incident is complete, dispatch closes the call for service, and the records management system is automatically updated for the appropriate fire department(s). These interfaces eliminate data entry duplication and have resulted in a FTE reassignment of 0.5 staff. There are various post-process tasks that consume the record data, such as false alarm billing and more advanced incident analysis.

Next steps

Now that the system is running for both the City of Ann Arbor and the City of Ypsilanti, the fire chiefs and HVA Regional Committee are actively pursuing deployment of this system countywide. This collaboration will assist in:

- strengthening local partnerships,
- having a positive financial impact, given new agencies must only consider hardware and software costs to "plug into" the existing system,

- funding for smaller agencies that would not have the budget or technical support to accomplish this on their own, and

- joint responses during a major emergency.

Figure 4 and Figure 5 depict this regional consortium at a high level:

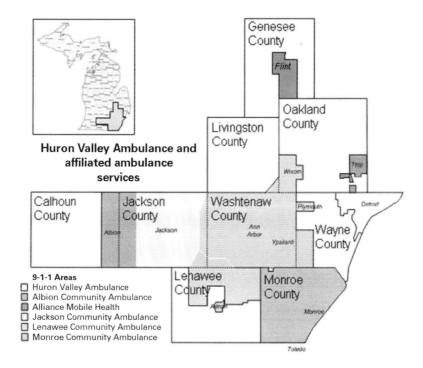

Figure 4. HVA service area map.

Keys to our success

Large change management initiatives require many things to go right and very few things to go wrong. This initiative started out as a local change management exercise focused on improving business processes for one organization by leveraging existing software purchases, and it has blossomed into a possible regional cooperative, creating an outcome where the residents of Washtenaw County, Michigan, will be able to sleep better at night knowing that the fire departments across the county are operating with not just the common goal of keeping them all safe from harm, but also a common operating picture. The

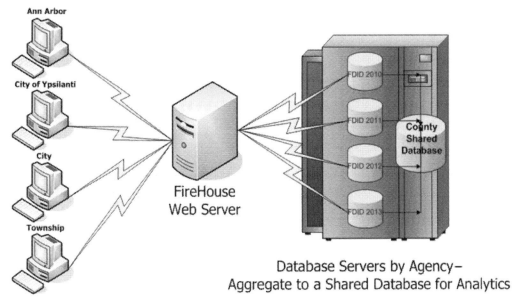

Figure 5. Hosted Solution Option: Hosted solution replaces individual RMS installations across the county, allows for an aggregate county shared database for analysis, and utilizes Web Server technology for access into the RMS system from virtually any system (desktop, mobile, phone.

road to success had, and still has, many challenges, including making sure that a well-defined business process could be developed and agreed to before any technology was changed or added, gaining acceptance at all levels of the organization, and ensuring that any new technology added or existing technology improved works flawlessly all the time. Overlaying all of this is the project leader, or leaders, and their ability to accurately assess the various customer audiences and customize the message of the day to meet the abilities of the room—for that day. Sometimes success may look like two steps forward and one step back, for a net gain of one step forward. This is still a win and needs to be celebrated!

The first thing that we did right was developed and agreed to a common business process before anybody changed anything. This did take a bit of planning on both the fire department and information technology teams' sides. First we had to develop awareness around business process and business process mapping with our customers, and then we applied that awareness to the existing work processes with the outcome being a high-level business process map. Sometimes, one of the unintended outcomes can be immediate incremental improvement in business process because an obvious bottleneck appears to the business in this process map, and a work process change can result in huge improvements. The business process map became a key piece of documentation that was used throughout the process by both the business and technology teams. It was particularly handy when Ann Arbor's fire department sat down with other fire departments and explained what was being accomplished. Sometimes, a picture is really is worth a 1,000 words!

Getting all the parties to buy into a large change management process was probably the most difficult, and most important, of all the tasks that needed to be completed. It is fairly easy to sell management, at most levels, on business improvements. An efficiency gain is the language of management, and so any improvements that will result in either reduction in time or reduction in cost are easily accepted at the management level. Acceptance of major change at the staff level can be more difficult for various reasons, including fear of the unknown, overall resistance to change, lack of trust for management doing the right thing, and a whole host of other reasons. The biggest obstacle to successful change is fear. Our experience showed us that making sure that everyone heard the same things, understood the why's, were listened to when they expressed concerns, and most importantly, fully understood how their roles were going to be impacted or not impacted by this change helped alleviate the fears that staff were experiencing. It is important to remember that this step takes lots of time, more than you would ever expect. Taking the time is one of the best investments you can make in ensuring a successful change management process.

Adding or changing technology often can be assumed to be the easiest part of any large change management process that involves investing in or improving existing technology. Not taking the time to carefully plan for the implementation as well as the continued operation of any of these new investments may be just what an uncommitted customer needs to derail what would otherwise be an existing project. This project's technology implementation and operation challenges specifically consisted of interfacing with a third-party CAD system not under control of Ann Arbor's information technology department, interfacing to two third-party off-the-shelf packages not under control of Ann Arbor's information technology department, and utilization of a local phone carrier's wireless data network as well as Global Positioning System (GPS). There are many points of failure in this configuration, and each one required significant time investment to fully understand the processes that need to be developed to ensure a high-availability production environment.

Having a project leader with a fairly well developed emotional intelligence is vital to the success of a complex project like this. Emotional intelligence is essentially a self-awareness on various planes of consciousness, including self-awareness, social awareness, relationship building, and self-management. A person with a well-developed emotional intelligence will understand the temperature of a room and the ability of the people to accept a certain degree of change or not. That person will change their delivery based on the situation. Additionally that person will understand their strengths and weaknesses, and find ways to leverage their strengths or strategically expose their weaknesses. A person with a well-developed sense of emotional intelligence also will understand the various stresses affecting a group and develop ways to address the "elephant in the room," sometimes even by leading the group to bring up those subjects themselves. This type of project leader will understand how to nudge, nudge, and nudge—and know when they can really push. Having a project manager like this will help make these complex projects successful.

Conclusion

The city of Ann Arbor's fire department and information technology department, along with their partners—the Huron Valley Ambulance Authority, Optimerix, Xerox, and Sprint—have combined forces to deliver a cost-effective and efficient combination of technology and process, resulting in a public safety technology service offering that can change the way this region manages its ever-decreasing emergency response fleet. By combining leadership's vision for change, the willingness to over communicate, effective change management techniques, and emotionally intelligent project management, the dream of a regional cooperative for fire and emergency response is now much closer to being a reality. Continuing to nudge the remaining interested agencies along the path of change still will take some time, but the authors are sure that with the proper amount of attention and focus, this cooperative will be viewed as a regional success story in interagency cooperation and collaboration.

Key participants

- City of Ann Arbor Fire and Information Technology departments
- Washtenaw County Administration and Information Technology departments
- Huron Valley Ambulance
- City of Ypsilanti Fire Department
- Optimetrics: Adashi software vendor
- Xerox FireHouse: Fire Records company/solution
- ESRI: Geographical Information System
- Other local Washtenaw County Fire Departments

DAN RAINEY serves as Information Technology Director and Chief Information Officer for the City of Ann Arbor, Michigan, where he has led the transformation of the city's IT department into an award winning, nationally recognized and respected service, most recently receiving its fourth consecutive Digital Cities Survey Award, given by the Center for Digital Government. In 2011, Dan was recognized as a Computerworld Premier 100 Leader, and in 2009 Dan received the Michigan Excellence in Technology Leadership Award. Dan holds a master's degree in Strategic Leadership from Walsh College and Business, and a Bachelor of Arts degree in Computer Science from Wayne State University.

JASON MCKINLEY serves as a Senior Application Specialist for the City of Ann Arbor and an Adjunct IT Professor for Cleary University, also located in Ann Arbor. Jason has more than a decade of management consulting experience within the government industry and a significant background in international IT project management. Jason has the ability to kill a yak from 200 yards away with his mind...just like the guy in Scanners. Jason has participated in numerous PTI functions over the past five years, including a PTI webinar detailing successful IT governance models. Jason holds a master's degree in Business Administration from Cleary University and a bachelor's degree in Urban Planning from Michigan State University.

Chapter 26: Integrated Public Safety Operational Control Centers

BY DANIEL PROCTOR AND DAVID KIPP

I. Introduction and overview

Public Safety Operations Centers (PSOCs) come in a bewildering variety of names and support an equally wide range of purposes. In some ways, the centers are as singular as the localities they serve, whether the locality is local government, regional multi-jurisdictional response, or institutional. But in all cases, public safety operations facilities share some common characteristics:

- Provide time-sensitive mission critical service for safety, security, and incident response
- Dependent on information and communication technologies
- Deal with multiple stakeholders, agencies, and jurisdictions—each with their own operational cultural and governance
- Complex and variable operational processes
- Knowledge-driven
- Variable levels of information access

The objective of this chapter is to provide a conceptual background for these facilities—a background compiled from recent advances in technologies and approach—and recommend methodologies for developing state-of-the-art operations centers.

PSOCs often are identified as command centers, communication centers, public safety centers, emergency operations centers, dispatch centers, 911 centers, public safety answering points, regional response centers or variations of those terms. The emphasis and taxonomy may vary, depending on the mission of the organization and the specific functions housed in the center, and understanding those functions is extremely important.

The common temptation is to view PSOCs through primary lenses of technology and facility. Indeed, almost any review of extant literature and research will focus on one of those two aspects. It is true that effective operations centers depend on specific adaption of building architecture and IT/communication technologies. However, the growing experience of public sector agencies is that even well-designed buildings equipped with reliable advanced technology will fail if the underlying operational processes are not adequately understood.

Common issues that emerge include jurisdiction disputes, inefficient incident response, poor communication/collaboration and lack of domain awareness, all of which in a public agency environment cause heightened concerns.

In this chapter, the concept of knowledge-based development of operations control centers will be used to establish recommended practices. Agencies are rapidly adopting holistic views of mission-critical facilities that acknowledge the pivotal roles of processes and knowledge.

Francis Bacon—16th century philosopher who coined the expression "Knowledge is power".

Knowledge is the currency of all public safety centers. Knowledge is information organized to do a job; it empowers the decision maker. It allows him or her to comprehend the operational situation sufficiently to take action and allocate appropriate resources. The timely production, formatting, and dissemination of knowledge is one of two principal components of the basis of design of centers (the other being capability—discussed later). If the planner fails to properly align the processes, procedures, forms, and technologies to the production of the knowledge required for watch standers to successfully execute mission, then there is no meaningful basis of inspection and testing to ensure that the center delivered meets the operational requirements of the end user. The emerging role of knowledge is explored in greater depth later in the development methodologies section of this chapter.

The chapter is organized in five sections:

1. Issues and advances

2. Functional characteristics and requirements

3. Technology requirements

4. Facility requirements

5. Integration

The order of the topics should not be misinterpreted as sequential. Rather, the experience of most public safety centers is that the most comprehensive understanding of a PSOC occurs when knowledge and capability are placed at the core of the facility, with process, technology, and facility supporting elements.

From the CIO perspective, it is imperative to adopt a view that is not conventionally IT-centric. Rather, a view of the facility that incorporates an understanding of processes leads to a more balanced—and effective—management strategy.

II. Issues and advances

Figure 1. Public Safety Operations Centers Depend on Three Key Elements.

The nature of PSOCs has changed dramatically in the 10 years following the events of September 11, 2001, and several catastrophic natural disasters such as Hurricane Katrina, the Alabama tornadoes, aircraft accidents, and even heat waves. The role of public safety in general has acquired considerably more prominence, and the role of operations, incident management, and response has vaulted to the forefront of public safety management issues. Facilities that have heretofore been the exclusive province of local law enforcement, focusing on call/dispatch/911 operations, have been altered in focus, function, and scale. The PSOC, in addition to traditional law enforcement operations, is increasingly seen and used as a nexus for the protection and recovery of critical national infrastructure. The important shifts in this area fall into four categories:

IT/security convergence

The digital revolution has been well documented with respect to security technology, and it is beyond the scope of this chapter to explore the technology of digital and network-based security systems. The result of rapid adoption of Ethernet and Transmission Control Protocol/Internet Protocol (TCP/IP) standards has infused information technology into the realm of security. Systems like closed circuit television, intrusion detection, radio, enhanced 911, and telephone have migrated to all digital architecture, software-based functionality and integration with other systems. As a result, the PSOC has, from a technology perspective, become an information and communications center, where the practices and applications of the IT industry more readily apply.

Multi-agency and multi-jurisdiction operations

Nearly every local PSOC supports police, fire, and emergency management agencies, as well as various mutual aid agencies in the event of incidents or emergencies. Increasingly, the facilities also deal with state and federal homeland security departments, state and federal transportation agencies, public works, utilities, operations and maintenance departments, and the FBI. The opportunities for jurisdictional conflicts and process/procedure conflicts increase accordingly, and become most apparent in difficult, high-pressure situations.

Critical infrastructure protection

The phrase "critical infrastructure protection" dates to the Clinton Administration and gained prominence in the aftermath of 9/11 in the United States. It underscores the importance certain sectors of infrastructure play in the economic well-being of a region or nation. Critical infrastructure includes:

- Banking and finance
- Electrical power
- Transportation
- Information and communications
- Federal and municipal services
- Emergency services
- Fire departments
- Law enforcement
- Public works

The impact of critical infrastructure protection on operations centers is significant. Methodologies, processes, and knowledge associated with these sectors become complex, increasing the reliance on systemic awareness and actionable information.

Infrastructure resiliency

The ability of a local or regional government subdivision to recover from natural or manmade events also has become a key aspect of the modern PSOC. As with critical infrastructure protection, the introduction of resiliency into the functional makeup of an operations center necessitates planning, information, knowledge, and response processes that have not been elements of traditional law enforcement facilities.

III. Functional characteristics and requirements

Identification of a PSOC's principal stakeholders is undeniably one of the most important factors in developing an effective center. It also can be one of the most challenging, as the stakeholders may include a number of agencies or departments that are not resident in the center itself but who rely on it or support it in various ways. The following stakeholder list is not meant to be comprehensive, as models vary widely, but representative and meant as a starting point.

Potential PSOC stakeholders

- Department of emergency management
- 911 dispatch
- Fire and emergency medical services department
- City/county police
- Public works
- Information technology
- State department of homeland security
- Emergency communications
- U.S. Coast Guard
- State department of transportation

Operations center leaders are well-advised to define the stakeholders and their operational requirements fully. Not doing so invites great risk.

Each stakeholder comes with a series of processes and procedures that are followed under defined conditions. Sometimes these processes are mapped and documented carefully, and other times they are not. In the latter case, it is not always necessary to map or document processes to gain a functional understanding of the requirements. Whether documented formally or not, the goal is to understand the critical processes, how information is acquired, processed and delivered, and action executed.

The following graphics (see next page) from the Federal Highway Administration demonstrate typical process flow diagrams for traffic and special event management, showing linkages.

CHAPTER 26: INTEGRATED PUBLIC SAFETY OPERATIONAL CONTROL CENTERS

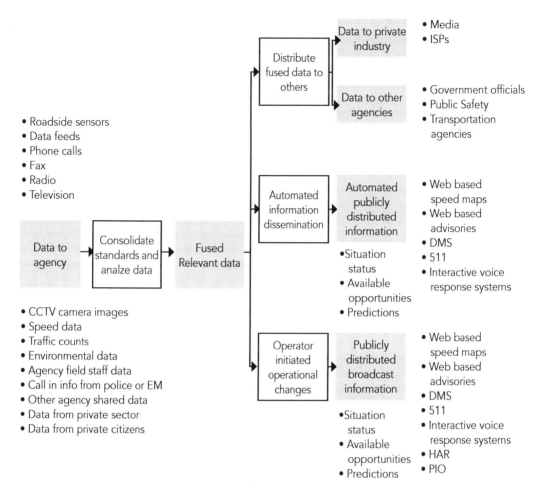

Figure 2.

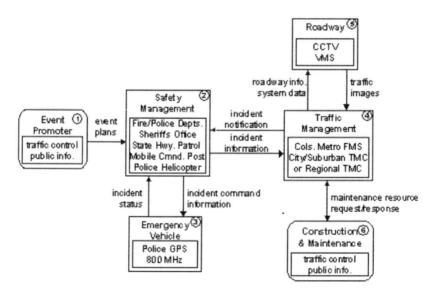

Figure 3. Special event management.

When the various processes of an operations center are considered, the complexity of the endeavor becomes apparent. Representative operational processes that may be found in a PSOC include:

- Emergency call reception requesting police, fire, medical, or other emergency services

- Assessment of emergency calls

- 911 emergency calls

- Monitor direct emergency alarms

- Information regarding vehicle registration, driving records and warrants, and provides pertinent data

- Incidents and related information such as nearest exposures and premise history

- Resource availability and/or capability

- Location of critical infrastructure and assets

- Roadway conditions such as traffic congestion and blocked routes

- Environmental conditions

- Determining value to the community such as power, communication, transportation, flood prevention, or basic necessities (food and water)

- Prioritizing values at risk based on loss significance (economics, health and welfare, cultural value, etc.)

- Assessing vulnerability and modeling potential events, infrastructure failures, damages, and other interdependencies

The sheer extent and complexity of processes that underlay a modern PSOC invites examination of effective information management strategies. This effort has led to emerging thought and practice regarding knowledge hubs or fusion centers, which will be taken up later in this chapter.

From the CIO perspective, the stakeholder and process underpinnings must be treated with utmost understanding, as this understanding leads to the rational conceptualization and deployment of technology.

IV. Technology requirements

The second primary component of an operations center is the suite of technology systems that provide communication, analysis, information, and computing to support the delivery of services defined by the required capabilities of the center. Like all information technologies, change is relatively rapid, with new introductions a constant feature of the horizon. The following paragraphs describe the most common systems required by an operations center and the current trends in those systems.

Voice communication

Voice communication in this chapter refers to telephony; radio communication is covered later in this section. In the operations center environment, telephony comes in two varieties: standard office telephone systems and emergency 911 systems.

Office telephony is moving steadily toward adoption of Voice over Internet Protocol (VoIP), although many facilities still deploy traditional Private Branch Exchange (PBX) telephone systems using PSTN networks. VoIP systems, which transmit voice signals on Internet-based communication networks, will become the norm. The inherent flexibility and mobility associated with VoIP, as it gains public popularity, poses particular challenges for public safety response organizations, which depend on determination of call location.

The introduction of IP telephony and widespread adoption of cellular telephones gave rise to the development of Enhanced 911 (E911). E911 is a technology that is designed to enable landline telephones to transmit critical location data quickly and transparently, as well as to provide accurate emergency-calling capabilities to mobile and VoIP phones. The system automatically connects callers to the closest Public Safety Answering Point (PSAP). Figure 4 depicts the basic information flow for E911 calls from enterprise telephone systems, in which the caller's physical location is transmitted to the correct PSAP.

Radio communication

Like digitalization of telephony and its attendant impact on public safety communication, radio is also in the midst of tectonic change. Along with advances in wireless digital communication, the desire for interoperable radio communication between users is also a significant driver of radio system changes. Technical discussion of radio frequency communication can be extremely complex and is well beyond the scope of this chapter. The focus here will remain on the impact radio systems have on public safety operations centers.

Conventional public safety radio systems operate in 10 different frequency bands, including the relatively new 4.9 GHz frequencies that are the subject of considerable debate. Radio interoperability is regularly named as a deficiency in emergency response. Radios that communicate on different frequencies and by different methods are very problematic in terms of cohesive and timely communication.

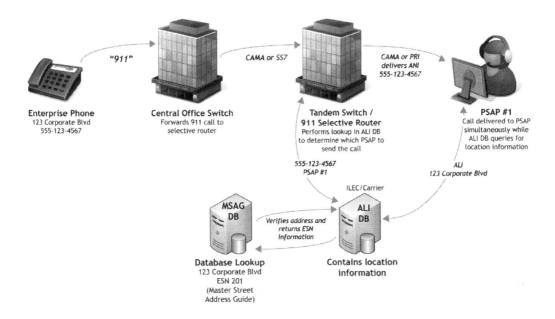

Figure 4.

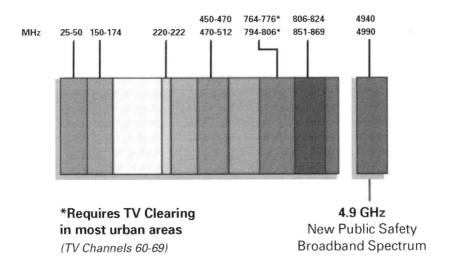

*Requires TV Clearing
in most urban areas
(TV Channels 60-69)

4.9 GHz
New Public Safety
Broadband Spectrum

Figure 5.

Proponents of "broadband" public safety radio communication cite the advantages of using IP networks as a means of providing seamless, flexible communication among disparate devices and systems. Many public safety professionals, even those who support Long Term Evolution (LTE) as a migration path for public safety radio, argue that without technical standards, high wireless network reliability, and ubiquitous network build out in place, conventional Land Mobile Radio systems will be widely used and expanded in the foreseeable future.

It is safe to assume that both approaches may be encountered. In a conventional narrow band radio system, integrated radio dispatch equipment—including base station/repeater, power supplies, interface units, and servers—will be present and generally supplied by the radio system vendor. In an IP radio environment, the equipment primarily will be standard networking devices and will be familiar to any IT organization.

The CIO should systematically review the totality of circumstances—including financial, operational, performance, reliability, and maintenance—in considering questions related to radio systems. The visible effect on the operations center will not be as dramatic as the invisible effects on staff training and operations.

PSIM platforms

Physical Security Information Management (PSIM) combines security sensor management with information processing and computing intelligence. PSIM systems are the next generation of Computer Aided Dispatch (CAD) systems that have been in use by public safety agencies for many years. PSIM platforms are designed to integrate and analyze information from physical security sensors and systems, and present relevant data to operators for resolution.

A typical PSIM software system provides all the necessary tools for situation management, including:

- Data collection
- Verification
- Analysis
- Resolution
- Tracking

All public safety operations centers will have PSIM or CAD software running on desktop workstations or consoles. Like many Internet-based applications, the trend will increase to aggregate platforms onto desktops, and older, equipment-specific built-in consoles will continue to diminish. Most operations centers will trend toward simpler multi-application workstations.

Introduction of more integrated PSIM systems result in consolidation of a large number of systems and applications that formerly required dedicated equipment, including government databases, mobile devices, security systems, and local/regional GIS maps.

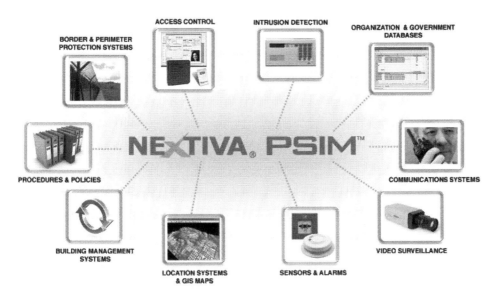

Figure 6.

V. Facility requirements

Although this is the starting point for many PSOC projects, the more progressive view is that facility requirements support the operational model adopted in the facility. Nonetheless, the facility requirements are critical to effective operation and include the following:

- Site layout
- Building architecture
- Structural
- Heating, ventilation and air conditioning
- Fire protection
- Electrical power
- Lighting
- Life safety
- Security

Key issues that arise in the operations center environment are staffing levels, staff comfort, ergonomics, and utility reliability. The following table (see next page) provides an overview of the important considerations that need to be addressed in a facility design or review.

Facility Element	Key Considerations
Site Design	parking, access, fencing, future development space, external equipment (radio towers)
Building Architecture	staffing levels, stakeholders, daylighting, dispatcher ergonomics, department adjacencies
Structural	wind loads, seismic loads, anti-terrorism force protection
HVAC	energy efficiency, redundancy, ventilation, comfort
Fire Protection	wet pipe/dry pipe/clean agent
Electrical Power	redundancy/reliability, generators, uninterruptible power supply
Lighting	interior lighting for operations area, office areas, emergency lighting, exterior lighting
Life Safety	fire alarm, code review, emergency mass notification
Security	building access control, CCTV, building blast resistance, interior zone security

Figure 7.

VI. Integration

The foregoing sections have concentrated on individual elements of a contemporary PSOC. The three most critical elements—processes, technologies, and facility—form the basis for an effective operations center. Yet, the three elements taken independently still do not guarantee that the center will consistently meet its missions and objectives. The remaining missing piece is integration—connecting process, technology, and facility into a coherent whole. This is the topic to which we turn.

Integration is a notoriously elusive term used in many situations to mean many things. In the context of PSOCs, integration will refer to the process of unifying operational processes, communication/information technologies, and the physical facility into a knowledge-based center capable of delivering the diverse, all-hazards detection, response, and recovery missions for which it is intended.

In the past five years, national public safety attention has justifiably moved from traditional public safety centers to "fusion centers" that carry a broader mission. Even in cases where the broader scope of a fusion center may seem inapplicable, the development of capabilities for prevention, protection, response, and recovery are applicable to even the narrowest interpretation of modern public safety.

According to the U.S. Department of Justice's (DOJ) Global Justice Information Sharing Initiative's (Global) *Fusion Center Guidelines*, "a fusion center will have the necessary structures, processes, and tools in place to support the gathering, processing, analysis, and dissemination of terrorism, homeland security, and law enforcement information."

The benefits and outcomes of adopting this holistic approach to development of PSOCs are described in the following table.

Strategic Activity	Tangible Benefit	Resulting Outcomes
Multijurisdictional/ multiagency collaboration.	Enables the entire fusion community to leverage resources from different agencies. Promotes efficiencies and enhances the potential for effectiveness.	Duplication of effort is reduced or eliminated. This equates to lower operational costs through a more efficient use of personnel and resources.
Create or enhance fusion center and partner information sharing in a manner consistent with other communities.	Enables future integration in an economical manner. Eliminates or reduces barriers that impede communication and intelligence development and exchange. Leverages current investments (across public safety and intelligence communities at each jurisdictional level) to offset cost and time. Affords opportunities for cross-agency and fusion center backup in situations where immediate resources are unavailable or insufficient.	Increased efficiencies in gathering critical investigative information. Analysts and investigators are able to access a broad range of information from a variety of sources, which lends itself to lowering the economic cost of crime through the sharing of common resources and the enhanced ability to prevent cross-jurisdictional crime.
Facilitate the processing and collation of disparate data.	Enables data captured from different sources to be organized in a manner that makes it useful for investigative and analytical purposes. Information is easily synthesized in support of tactical initiatives.	The ability to aggregate data from a variety of sources provides a broader foundation for analysis and decision making; better analysis through access to more comprehensive data should result in higher rates of detection, clearances, and prevention of multijurisdictional crime and potential terrorist threats.
Application of analysis techniques enabled by advanced analytical tools.	Enables analytical services, such as crime pattern analysis, association analysis, telephone-toll analysis, flowcharting, financial analysis, and strategic analysis. Analysts are able to describe, understand, and map criminality and the criminal business process. Provides information to engage the most appropriate tactics.	Enables informed decision and offers better choices for decision makers. Enables the targeted and better use of resources, thus enhancing efficiencies. Disrupts prolific criminals and terrorist threats through targeted decision making.

Figure 8.

Development methodologies

The design for all centers starts and ends with the planners' clear understanding of the mission of the center—an understanding that is organized in terms of the knowledge and capability the center must provide watch-standers so they can best execute mission. Form, processes, and technologies that create the center must be guided by the knowledge watch-standers need to acquire and maintain of their operational environment and by the capabilities or tools required by those officers to manage that knowledge and disseminate it to best mitigate public safety concerns with the most effective allocation of scarce resources.

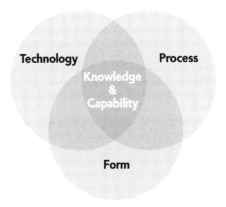

Figure 8.

At the core of this concept is the formation and dissemination of relevant knowledge. Relevant knowledge is more than information or even actionable intelligence; it is acquired and organized information relevant to accomplish a task, delivered to the proper person at the proper time.

Generating and documenting knowledge requirements

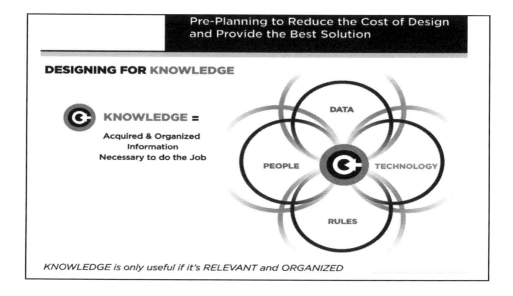

Figure 10.

The path to developing knowledge requirements involves developing a deep understanding of the operations center mission and the specific knowledge the decision makers must acquire from their operational environment to best execute their mission. This information is usually generated from a close examination of the public safety or security concerns that the center was envisioned to mitigate and manage. This information must be collected and documented in sufficient detail that it can inform all subsequent design milestones—from the conceptual design documents to the final testing and commissioning—of systems that comprise the center.

For small centers created for single public safety or security organizations, the process to generate and document knowledge requirements is relatively straight forward. However, for large centers where many independent organizations gather to manage events globally, such as an Emergency Operations Center or a Port Command Center, the process is more complex—organizational rivalries, cultural barriers, and economic competition can quickly impede even the most organized effort to develop global knowledge requirements.

Yet, determining those requirements are imperative for the future success of the organization. Most planners will tackle this challenge by having their technical professionals collaborate to create center-wide requirements, but this approach, while common, is fraught with challenges: the process creates technically specific requirements that are poorly linked to the operational requirements of the end users, and the requirements created reflect the needs of individual organizations acting independently instead of the needs of all organizations collaborating together to address concerns that will require their combined actions.

The more effective approach to developing center knowledge requirements is to enlist the aid of external consultants to facilitate the process and document the results. Using external consultants is more efficient because they are trained in facilitation techniques, have tools and processes already developed for tasks, and most importantly, they are not vested in any one of the organizations that will populate the completed center. This latter point is significant. Experience has demonstrated that when a requirements development process is attempted by one of the center stakeholders, the other center stakeholders are disinclined to participate meaningfully—they are concerned that doing so will enhance the role of the stakeholder organization charged with the facilitation at their expense. An external consultant is viewed as an unbiased agent; consequently, stakeholders are usually more open and collaborative with the consultant.

Once collected, knowledge requirements must then be examined from the lens of the center's concept of operations (or CONOPs), and processes and procedures, and then parsed and tailored to create system-of-system performance (or functional) requirements. These, in turn, inform the creation of equipment specifications for the center's systems. The totality of this data must be captured in a traceability matrix, which will subsequently steer system testing and commissioning.

Identify future capabilities required by the center

In addition to providing decision makers the knowledge they need to comprehend their operational environment, the center must be designed to provide watch officers the ability to manage the knowledge created, to share that knowledge with external stakeholders and collaborate, and to coordinate the activities of all resources that have a role in the mission of the center. The process to define the capabilities of the center considers the mission of the organization, the seminal public safety and security concerns that influenced the creation of the center, and the operational process and procedures of the center.

Again, when planning the capabilities of large centers with many different jurisdictional partners, the planning process is well-served by embracing the value of external consultants to facilitate the effort. Doing so enhances the efficiency of planning and promotes greater "out-of-the-box" contemplation and creativity in the planning team, as well as mitigating against the propensity for "silo planning" by individual agency stakeholders. And, as with knowledge planning, the capability planning approach must start and end with the operational needs of the end users—the individuals charged with managing operations and events. Clearly, there are many stakeholders in the center, and all of their requirements must be considered, but operational expediency of the center should always trump other planning considerations.

When the functions and processes are understood in depth, a surprising picture emerges of the areas and types of knowledge required for a multi-jurisdictional operations center. The figure below, taken from the U.S. Department of Homeland Security's National Preparedness Guidelines, demonstrates the effort to define relevant knowledge.

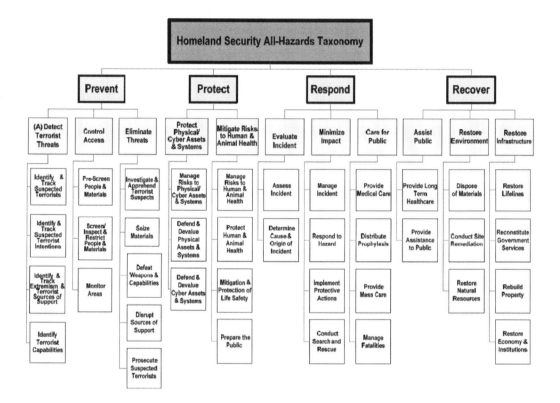

Figure 11.

Strongly associated with knowledge-based design is the development of a concept of operations (CONOPS), which is increasingly a required document in the development of an integrated public safety communications center. As in the development of operational process requirements, the notion of a CONOPS is to capture and develop the integrated operations from the perspective of the users. The reason that CONOPS have become more prevalent is that operations centers have engaged more complex missions, requiring specialized communications with some level of integration required to establish domain awareness based on knowledge.

With CONOPS and knowledge requirements defined, an integrated PSOC can be developed with appropriate technology and facility elements in place to support processes and information sharing. Many such facilities are being developed in state and local government agencies throughout the United States.

Figure 12.

VII. Conclusion

Critical operations centers are growing in scope and complexity as the world adjusts to the realities of the 21st century. Historically, centers have been planned piece-meal by individual agencies identifying their needs, and then the planning team attempts to integrate those needs with the needs of other agency stakeholders. Centers were designed with an eye toward system and space configuration, replicating the style of the participating agency's own environment with little consideration for collaborative operations. Experience has shown that such an approach yields unsatisfactory results. Another approach is to hand the planning off to a team of technologists from each agency stakeholder to envision the capabilities the center will need to support joint operations. Again, the results are unsatisfactory—systems are built and integrated and then not used because they don't provide content of value.

The most satisfactory approach is to design centers organically in a holistic process that considers form, processes, and technology after first conducting a detailed assessment of the *knowledge* that the centers must provide the end users, what *content* must the systems provide the watch officers so they can best execute, and a similar assessment of the *capabilities* the facility must provide. This approach is organized to ensure that the design is informed by the needs of the end users working collaboratively to mitigate threats or potential threats in a coordinated fashion. It ensures that the center provides users actionable intelligence and the tools necessary to best manage joint resources. And, it ensures that all aspects of the center—the form, the rules and policies, and the technologies—are considered and planned in a coherent and integrated methodology.

DAVID A. KIPP, P.E., is Senior Vice President and partner at Ross & Baruzzini and provides executive leadership for the firm's technology consulting practice areas and global engineering assignments. Kipp has a distinguished record of advising public sector clients in the aviation, maritime, municipal government, education, and healthcare industries. Recent focus areas include critical infrastructure protection, information technology strategic planning, mission-critical facility development, and wireless applications. He also serves as the project director for a number of large technology infrastructure development projects and is also a recognized contributor to the fields of aviation technology and public security. His published articles and presentations have appeared in many national transportation and government forums.

DANIEL PROCTOR, Vice President, Ross & Baruzzini, specializes in developing situational awareness solutions for the transportation, government, and commercial security markets. With more than 25 years of experience in the global security industry, he has extensive experience reengineering, consolidating, and optimizing business units and technologies in the security arena, especially business units that are judged by the quality of information they contribute to organizational effectiveness in risk mitigation.

Chapter 27: Interoperable Communications in Emergency Management

BY EVERETT L. DAVIS

The Lubbock County, Texas, Mobile Operations Vehicle (MOV) provides Internet access, data, wired and wireless Voice over IP using existing phone lines from the core network. Utilizing inter-operable radio communications, including Wi-Fi, it has been repeatedly tested in exercises and proven in events.

It also serves the State of Texas Rapid Response Task Force as a statewide asset. Resilience of the Lubbock County data network supports Voice over IP telephony and data services, making the MOV a mobile office building inheriting operational sustainability of the core network. Utilizing VoIP through satellite, it can deliver phone and fax services, avoiding traditional SAT phone costs, to a MOV deployable anywhere in North America.

Recent additions of two smaller MOV units in the West Texas jurisdictions of the City of Plainview and Garza County are integrated into the VoIP phone network at Lubbock County. This enables the three units to simultaneously deploy in different areas of the state and to inter-operate using the common infrastructure at Lubbock County. Configuration is poised to support additional jurisdictions that may wish to develop similar capabilities at reduced cost.

Whether tornadoes, wildfires, hurricanes, etc., threaten the area, the units stand ready and the teams are trained to act when disaster strikes. The platform and structure, whether for

Figure 1.

mobile or stationary deployments, are helping set the standards for the state for functional interoperability, resilience, and jurisdictional continuity of operations.

The challenge

When small jurisdictions are hundreds of miles and hours away from major metropolitan areas of the state, local jurisdictions must work with each other to provide resilience and response to the residents they serve in creative ways, and nearly always without the financial resources and assets available to larger communities. After Hurricane Katrina, Lubbock County Emergency Teams deployed to support relief efforts. Action reports from the teams clearly illustrated the needs in West Texas and the South Plains for survivable sustainable communications, which were not present initially in the Katrina response. It was clear, if you do not take it with you, you will not have it at the time it's needed most, wherever you deploy.

The response

Lubbock County, Texas, had a vision that the South Plains Association of Governments embraced with Homeland Security Grant funding to equip a Regional Emergency Response Mobile Operations Vehicle that could support the 15 county region in Texas extending from the New Mexico border on the west to the Oklahoma border on the east by providing multi-jurisdictional interoperable radio, voice, video, and Internet access. Lubbock County realized that as the largest jurisdiction in the region, it already could

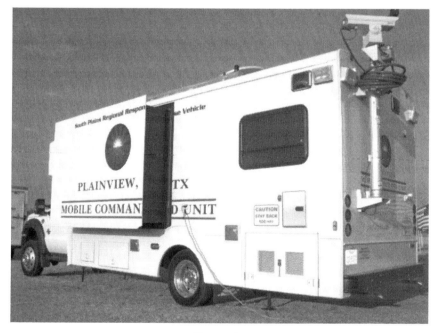

Figure 2.

provide the core infrastructure for support of both voice and data, utilizing its own highly resilient data network design as the interface backbone to support the mission, and allow other jurisdictions to interconnect and inherit the resilience of the core in the future.

The Lubbock County MOV, initially equipped with the Raytheon Tactical ACU-T Tactical Interconnect System for multi-jurisdictional interoperable radio, also was equipped with Cisco Voice over IP (VoIP) phones, connected via satellite dish and a Virtual Private Network to the Lubbock County Data Center, and dual Cisco Call Manager devices to its own existing telephone lines. Local, long-distance, and station-to-station intercom calls were now possible from and to the MOV, deployable anywhere in North America.

In February 2012, after approving inter-local agreements between Garza County and Plainview, Texas, Lubbock County proved the "Future is Now" by successfully interconnecting two smaller MOVs with Cisco VoIP phones, using different satellite service providers and Cisco VPNs, to the Lubbock County VoIP network core. They are positioned to interconnect by VoIP the remaining agencies in West Texas using either mobile or stationary structure-based assets as needed.

Results

The system has exceeded all expectations in deployments across the region and the state. It has been used in tornado responses, wildfire responses, onsite hostage negotiations, accident investigations, virtually every conceivable law enforcement tactical application, and has on more than one occasion deployed to provide incoming 911 call center answering services when cities in the region were isolated by severe fiber cuts. Area 911 calls were re-routed to numbers in the MOV until repairs were made and service restored.

In a Texas readiness exercise in Waco, the unit and team proved to be the first asset at the exercise that was operational and which always achieved and delivered the full range of services within its stated capabilities, resulting in the Lubbock County MOV being assigned as the primary vehicle of its type to the Texas Rapid Response Task Force for all hazards response across the state, and will be requested to pre-stage in Waco, Texas, for hurricane response events.

Previous SAT phone costs for 24 hours exceed the standard cost for the standard five-day monthly service for the MOV. Billing three-hour service intervals allows the unit to deploy for very brief events with no cost penalty with far greater functionality.

How did we get where we are today?

Lubbock County began many years ago focusing on our defined statutory and adopted missions, realizing that continuity of governance and resiliency depended on a well-defined inter-operable sustainable infrastructure that both supported the collective missions of over 40 departments and avoided duplicative technology silos, reducing hardware investments, support, and training costs to support them.

Sufficient resiliency

Lubbock County did not focus on resiliency solely to support emergency operations; instead, it focused on resiliency to support agency operations of 40 departments, which must equally be able to function when bad things happen. As a criminal justice agency, Lubbock County realized an integrated criminal justice management system simplified the complexities introduced with integrations of multiple disparate software vendors.

Lubbock County recognized the need for manageable system storage growth outside the typical server platform and was an early adopter of SAN storage. It is now in the third generation of SAN hardware, while forecasting the fourth. Lubbock County further recognized the structural and operational benefits of server virtualization, coupling it with SAN storage and auto-failover/auto-recovery/sustainability, initially envisioned in a single datacenter environment. We strove to get there.

A second data center

Built to replace the aging jail, Lubbock County constructed a new detention facility with a second datacenter located several miles from the courthouse and existing jail complex in the downtown campus. Housed in a hardened facility with dedicated divergent path underground fiber between sites, Lubbock County provided virtual fault tolerant switch pairs at each site and replicated SAN storage site to site, leveraging those same virtualization services to provide site-to-site resiliency and failover, initially providing 40 Gigabit site-to-site fiber interconnections.

Resilient Internet and PRI telephony

Dual site datacenters and network platform provides Lubbock County the ability to leverage dual provisioning of both Internet service delivery, and telephony services with Hot Standby Router Protocol provided by the ISP to sustain operations automatically should an event disrupt either service delivery or a hardware failure occur. Satellite VoIP MOVs or other externally connected devices for agencies benefit from resiliency in the hosting core as they attach.

Build it and plug it in

Lubbock County internally developed a video visitation platform for the new (remote) detention facility using over 260 diskless workstations in secure housings, with USB cameras while booting from an image on redundant servers. A default running application leveraged Adobe Flash server tools to make the connections and interfaces with recording-on-demand tools. The core network infrastructure was sufficient to the task and now offers video arraignment, secure (non-recorded or monitored) attorney visitation, etc., negating the need to move inmates to and from the courthouse for many processes.

Figure 3.

Security resilience

Firewall systems at each site equally provide protection and service continuity that auto sense disruptions and re-route so efficiently that active VPN connections, such as those supporting the MOV platforms or remote users, often are sustained through the disruptive event without VPN connection loss.

Often overlooked

Sites with levels of resiliency often fail to meet objectives as detection of the first event goes unnoticed. Resiliency masks the event from the users who never experienced a service disruption and may have no indication a failure has occurred.

It is essential to have a properly configured detection system to actively monitor critical elements, systems and communications segments, or a failure of a primary system component will remain undetected until a supplemental failure occurs weeks or even months later in the secondary, causing service disruption to users and exacerbating diagnostics of two events that appeared to manifest as one event, perhaps with some resulting data corruption.

It is essential to configure a monitoring and reporting system to provide active alerts to the helpdesk or network operations center on infrastructure component failures as they occur. Routine maintenance during normal business hours will reduce costs while increasing availability of domestically positioned vendor technical support groups for extended support issues that might arise.

It is all about the mission

In emergency management and communications, during a disaster is when the residents need the government agencies that support them to be providing services at a high level—

not to be consuming services to restore their own operations. Regrettably, fault tolerance was never meant to infer that we have tolerated these faults for years, yet it often does.

Procurement is only a component

What goes into the procurement and product selection process defines the ability one has to design, specify, procure, manage, upgrade, replace, and transition to the next generation of technology. In a disaster, the most successful manufacturers and industry leaders have statistically the larger base of knowledgeable support staff both internally and externally in the marketplace to assist you, and they have the breadth both physically and fiscally to help restore operations when disaster strikes.

Considered by some to be "The Common Law of Business Balance" is the statement often attributed to John Ruskin (though contested): "It's unwise to pay too much, but it's worse to pay too little...." It is worth consideration in building fault tolerant and resilient systems to sustain public sector and private sector alike.

Natural growth

Sufficiently resilient system foundational infrastructure should allow logical expansion with the resulting additions to almost effortlessly inherit the full resilience of the core. Long-term strategic planning with the involvement of those stakeholders who are most directly involved with achieving the mission for their department, agency, or company is a fundamental tenet of project management. Our agency has been focused on planning for years to come; it is an essential part of our mission.

COOP—Everyone needs it

Lubbock County engaged a professional firm in continuity of operations planning (COOP) for our agency and have addressed the needs we could foresee for recovery and sustainability, including primary, secondary, and tertiary locations to reestablish operations should physical damage from a disaster destroy one of our facilities, or if we simply require temporary relocation of a department for facility repair. COOP plans provide the framework and blueprint for our staff or their successors to logically function when the unthinkable actually happens, and lay out the processes to recover from lesser disruptions.

Are we truly prepared?

COOP is incomplete unless it includes the same COOP for the agency's staff and their family's resilience. It is the agency or business staff one relies upon to respond when disaster strikes and to carry out the COOP itself. If they cannot respond for whatever reason, the result is the same; there are plans, with insufficient staff to implement them.

How prepared are the vendors who will support the agency or business, and how well prepared are their staff to respond? Those questions are frequently overlooked in COOP. If one relies on their vendors' availability to restore affected services, it is crucial to include their abilities to respond in COOP.

Mission accomplished?

Does Lubbock County always hit the center of the bull's-eye of the target? No. All eyes are on the target, however, and we strive to hit it in the most efficient manner possible. Lubbock County worked together at many levels to both define and achieve what we have placed to date and what is yet to come. Many very fine and extremely dedicated people have contributed to the goals over many years.

Multi-agency coordination

By reaching out and working through the South Plains Association of Governments (SPAG), involving the surrounding jurisdictions who face the same challenges, we train side by side, conduct test exercises together, respond to incidents together, and have successfully managed large events successfully.

Vision for new solution sets out to save lives

On September 11, 2001, lives lost due to communication problems focused our attention on the fact that interoperable emergency communication capabilities were woefully inadequate locally, regionally, and nationally. Many funding initiatives and grant programs, such as P25 compatibility requirements, have subsequently worked to help address that problem, yet much remains to be done.

Our collective mission to save lives in a disaster where lives are at extreme risk has yet to address the communication issues of trapped victims armed solely with cell phones when all supporting infrastructure is destroyed or inoperable. Regulations exist that force all cellular systems to accept 911 calls from phones from any cell phone whether activated with a carrier or not.

Hurricane and tornado event hot washes and after-action reports are replete with instances where injured residents frantically call 911 with no surviving infrastructure to answer their calls. The phone batteries may deplete with no hope of reaching help before it is too late.

Law enforcement and the emergency response community can now speak with each other, yet often cannot speak with victims seeking their help, relying on search-and-rescue efforts to discover who is injured, trapped, or somehow remain in harm's way.

MOVs need a lightweight affordable mechanism to establish a (911 only) cellular response to the victims they deploy to serve. Responding communications radio staff must be able to both reach the victims and dispatch the appropriate personnel to assist. It will save lives.

Having researched this issue for several years and consulted with numerous industry leaders, we still have to ask the question, "Which companies are going to structure and provide a rapidly deployable 911 cellular technology to the emergency management communities?"

What might come next for Lubbock County and the surrounding region?

The supporting infrastructures are used actively in continuous operations by Lubbock County, thus their operational readiness benefits our agency and our neighbors who elect to partner with us. As Lubbock County continues to work with the South Plains Association of Governments, we have in place (with their significant help) a framework to attach nearly any kind of device one might envision. Connectivity could be as simple as a VoIP phone with VPN capabilities, video phones, interoperable video conferencing equipment, radio over IP, etc., in each jurisdiction in our region who choose to participate.

Connections for tablets, smart phones, additional MOVs or "all the above" are possible. The key is the availability of the centrally designed resilience of the core capable of supporting missions we can presently envision without having to replicate silos of expensive technology in every jurisdiction in the region. This is an example of local tax dollars at work. It certainly is not the work of one man, one department, or one agency. Many dedicated public servants and vendors have worked tirelessly together to make it possible.

Connecting new technology often can be as simple as acquiring a new type device and installing the license file to enable it. Expanding our reach could include connectivity with our regional neighbors to the north in the Texas Panhandle Region, and connectivity to the south with our regional neighbors in the Permian Basin Region, who each have achieved excellence in their own respective regional preparedness initiatives.

What comes next is truly up to you, in your agency, your business, your family, and your community.

Author's personal philosophy

A well-known humanitarian, the late Cal Farley, a man I knew well and greatly admired often said, "It's not where you have been, it is where you are going that counts." It has framed my life's work.

EVERETT L. DAVIS presently serves as the Lubbock County Manager of Network Planning & Support, as Lubbock County Technology Planning Director, on the Local Emergency Planning Committee, and the Regional Emergency Planning Committee. He serves as the Regional Technical Advisor to South Plains Association of Governments and is part of the Regional MACC team in West Texas, having trained at the U.S. Center for Domestic Preparedness, and the National Emergency Response and Rescue Training

Centers with a number of his Lubbock County and West Texas colleagues. He has served on the board of directors for the Texas Association of Government Information Technology Managers, and presently serves on the board of directors for Cal Farley's Boys Ranch Alumni Association. Mr. Davis is also a certified government CIO.

Chapter 28: Critical Components for Survivable, Sustainable, and Flexible Communication Centers

BY IAN REEVES

Over the past 40 years, Architects Design Group (ADG) has had the opportunity to be involved in the programming, master planning, and design of hardened facilities throughout the United States. It has provided ADG with an insight into the issues that are very specific to these facilities that must be survivable. Communications centers clearly are at the forefront of the nation's critical infrastructure, and as such, must be developed in such a manner as to ensure that they remain operational through an overlapping fabric of essential considerations.

There are significant regional influences on the design considerations for communications centers. They range from seismic activity, hurricane-force winds, tornadoes, ice storms, and flooding, as FEMA would refer to as "all hazards." These conditions must be driving factors in the planning, design, and engineering of these critical facilities.

Likewise, man-made disasters have become an unfortunate reality of this era. Basic site planning models also should be implemented to mitigate potential threats in our more urban environments, mirroring the three-tiered system used by the Urban Area Security Initiative (UASI) program, which has established the locations of the greatest potential threats through vulnerability assessments. Similarly, there are site development protocols that address site security considerations that are cost-effective and rather simple to implement. These are available through the International Crime Prevention through Environmental Design (CPTED) Association, and have become widely adopted by architects and planners that specialize in the field of public safety architecture.

Relative to the aspects of proper planning, design, and engineering of communications centers, there are distinctive components that are essential to the success of the facility development, which include survivability, flexibility, and redundancies. Each are equally important in that these facilities typically are designed with a 40- to 50-year life cycle, so the preliminary planning and subsequent facility design must be successful to not only meet the agencies' known current and future needs, but they must also accommodate the emerging trends, such as consolidation, technology advancements, and enhanced building code requirements.

Sustainable design is another relevant consideration, as this movement continues to become increasingly more prevalent in the requirements of governmental projects.

Figure 1.

Leadership in Energy and Environmental Design (LEED) certification of buildings has surpassed a grassroots effort to become widely adopted, as it promotes sustainable design and development practices by utilizing rating systems that qualify the various objectives accomplished in efforts to promote better environmental and health performance.

The future expansion of services provided from our nation's communications centers is very likely going to involve the inclusion of regionalized fusion centers. The real-time, crime-solving capabilities are being explored in a variety of applications nationwide. The communications centers are often the primary locations from which the traffic management and security camera networks within a community are monitored. The opportunities to expand this capability and to manage the data developed by the network into usable information in efforts to provide a higher level of public safety are seemingly endless. It is anticipated that this approach that is currently being tested shortly will become standard policy on a national basis in the development considerations of communications center projects serving America's communities.

Hardened facilities: Lessons learned—The protective envelope

The primary factor for consideration is the premise of facility survivability. For example, if the forces of a storm can pierce the protective enclosure of a facility, it will not, in almost all instances, survive as a functional environment. Thus, the primary goal should be to adopt the premise that the building should be capable of functioning regardless of the forces to which it is subjected.

As an example, hurricanes can be both unforgiving and unique in their ability to find the "weak link" in a protective envelope. A misplaced fresh air vent, as an example, when subjected to a Category-5 hurricane with wind forces of 156+ mph, quickly can be destroyed as exterior wind forces become interior wind forces.

The importance of a protective building envelope, in which every surface and particularly every opening has been critically examined, cannot be understated. Doors, windows, fresh air intake vents, roof vents, exhaust air vents, and service access panels are just a few of the normal penetrations that exist on buildings. In many cases, they represent points of access for systems or are code requirements that are mandated for life safety reasons. What, then, is of paramount importance is to review each and every such "opening" and to ensure that they are either designed to resist the forces generated by a storm, or that laboratory-tested protective elements can be applied easily to provide a secured envelope.

Nothing should be left to chance in this respect, as the building envelope is only as strong as its weakest element.

Exterior wall protection

The building structure itself also must ensure survivability, including the two primary elements: the walls and the roof. Of these, the wall system is more easily resolvable, being a product of appropriate architectural and structural design that considers both wind loading and the ability to resist the impact of airborne projectiles. A street sign post, when airborne and in a horizontal position, can easily penetrate most building wall types that have not been appropriately designed. Thus, the wall construction of the building becomes an important element in survivability.

Utilizing poured-in-place or cast-on-site (Tilt Wall Construction), high-strength and reinforced concrete is the material of choice, as it provides a strong and dense surface resistant to most storm forces. Concrete block or masonry units, when appropriately strengthened with steel reinforcing rods and filled solid with high-strength concrete also can provide protection but to a lesser degree than a monolithic poured-in-place concrete system. It is also of critical importance to ensure that the building design provides for lateral support of exterior walls, particularly those of larger spans. In essence, the ability to design appropriate exterior wall systems is also a product of the configuration and location of reinforced walls in the building's interior, a factor that must be taken into consideration as the functional characteristics of interior spaces are designed.

The roof: The weakest link

The roof of a building is, in many respects, the most difficult element of the exterior envelope to design in such a manner as to be able to resist the forces of wind and water. The experiences of recent storms have clearly shown that the point where the roofing material connects with the exterior wall is the area of critical importance. Once this "connection" fails, the roof in its entirety begins the process of destruction and inevitably leads to building failure.

There are specialized roofing systems that are appropriate for "hardened" facilities, generally being those that are applied to the roof decking in a series of adhered layers with mechanical fasteners. As the roofing elements get closer to the roof/wall connection, the magnitude of these fasteners increases, and the spacing of the fasteners decreases, providing additional protection against the destructive negative pressure uplift forces that are known to be most destructive.

It is highly recommended to utilize an architectural roofing specialist to assist with the design and proper specifying of a roof system that exceeds the building code criteria for the region in which a project is being planned. This will ensure that the most common weak link is appropriately addressed and does not become the prime candidate for facility failure. It is also important to note that the design of the sub-roof and roofing insulation must incorporate more restrictive standards for adhesion than is normally provided. The roofing composition must, in essence, be considered as an entire system and not as individual components.

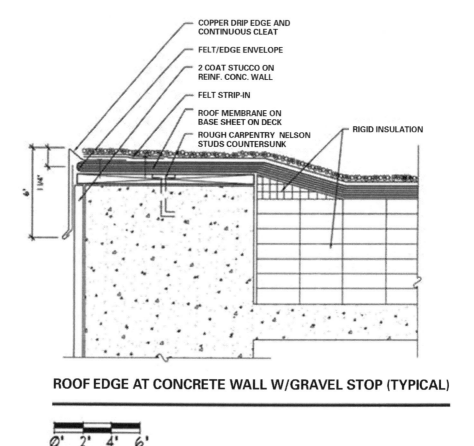

ROOF EDGE AT CONCRETE WALL W/GRAVEL STOP (TYPICAL)

Figure 2.

Glass: An integral component of the protective envelope

The premise of having a window to the outside world generally strikes fear in the heart of a director of the agencies that reside in hardened facilities. Historically, glass in either windows or doors has been perceived as a point of building vulnerability, easily destroyed in a storm.

The technology has changed as the industry has made significant advances in response to other needs and circumstances. Prisons, as an example, have evolved their management philosophy to now provide maximum visual access of activity. In order to accomplish this need, new formulations of glass and related products have been invented, tested, and utilized. They generally consist of multiple layers of tempered glass, a poly-carbon inner core, and other products that resist impact. These systems can resist a .45-caliber bullet fired at point-blank range without piercing the protective layers.

Astute designers have successfully introduced this technology into the world of "hardened" facility design. These facilities, by their nature, can be relatively stressful environ-

Figure 3.

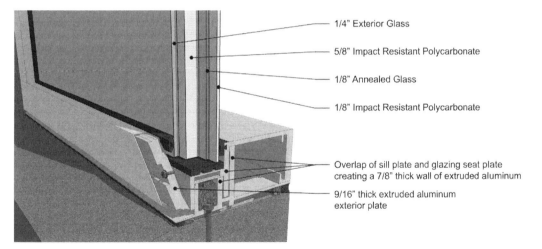

Figure 4.

ments. Most accomplished designers are strong believers of the psychological need of people to have work environments that have natural light and that offer the opportunity to see the outside world.

In terms of client acceptance, there has been some success, but certainly, it has not yet been universally accepted. There are very specific standards that must be met to ensure the appropriate level of protection with such systems, but there exists sufficient documentation to support its serious consideration. It is strongly recommend that any existing or proposed facility consider the use of these types of systems, as they will enhance the work space while providing a secure and protected environment.

Flexible technology

Of all of the components that are involved in the development of a communications center, technology is the one that continues to evolve most frequently. One could surmise that within two years of a new facility becoming operational, the technology will have advanced, potentially requiring facility upgrades relative to specialized cooling and ventilation systems to address the additional equipment that may be part of the technology equipment upgrades. As more equipment is added to the interior of the structure, the higher the heat gain is from the venting of the equipment and, thus, the added cooling loads to maintain the optimum room temperature of the areas housing this specialized equipment.

Architects and engineers familiar with the nuances of properly planning and designing communications centers understand the necessity for flexibility in meeting the ongoing demands for emerging technology. The inclusion of open raceways, accessible cable trays, and "miles" of open conduit with pull strings for future considerations are a must. The additional conduit should occur not just within the structure but also from the exterior to the interior, such as at the point of connectivity to a communications tower location.

The electrical engineer's design for any planned facility should include electrical panels with multiple extra breakers in each panel for an even distribution within the structure to accommodate future electrical loads of new equipment. The dual electrical services coming into the building should both provide a minimum of 20-25% additional capacity to what the known loads of the planned project will require at the time the facility becomes operational. This minor increase in development costs will more than pay for itself over the 40- to 50-year lifecycle of the facility.

Critical systems infrastructure redundancies

The issue of systems redundancy cannot be understated, as it directly relates to the premise of building survivability. It should be accepted as a given that public safety/communications/law enforcement/EOC facilities must be self-sufficient in that they should be capable of continuing operations when most, if not all, normal services to and from a building have failed for a minimum of 72 hours at full operational capacity.

Sewer systems, water supply, power, and communications are all equally important elements, and the failure of any one can, and in all probability will, result in facility failure. Sewer systems, as an example, may be provided by a forced main in which the transmission of sewage is made possible by off-site pump stations. The failure of an electrical supply system can result in the failure of these pumps and cause a back-up in the system. Such an event quickly results in facility failure.

The loss of normal emergency power and its effect on operations has long been understood. What has not been clearly comprehended is that providing a singular emergency generator does not ensure survivability. Access to storage of fuel (diesel or natural gas) and lubricating oil are both needed. When a facility is engineered to use multiple generators, concrete blast baffles should be constructed between each of the generators to ensure that a major mechanical failure in one generator does not compromise the unit adjacent to it, as has been documented to occur.

What must be adopted is the premise of redundancy, meaning that each critical building service or system also be provided with a back-up capability. Any new facility of this type should consider the incorporation of an on-site septic tank system in the event of failure of the basic sewer system, and the ability to cut off basic systems to avoid a back-up that contaminates the facility.

The facility also should be provided with either its own emergency potable water system, such as a well in a protected environment, or by using inflatable bladders in which water can be stored in a protected environment prior to an event. Chlorination tanks are required for most potable water wells for drinking purposes. This equipment requires space within the structure and should be clearly noted in the pre-design planning efforts of any proposed project. Just as one should critically review the building protective exterior envelope, one should develop and implement redundant systems that are critical to facility survivability.

Figure 5.

Figure 6.

Sustainability (LEED)

Architects, developers, construction professionals, and clients, are being called upon to promote responsible building and design with respect for the past and an eye toward the present and the future. The success of sustainable architecture is the integration of style and substance. It is important to be able to demonstrate how building "green" does not mean you have to camouflage the facility within the landscape. Instead, a structure can harmonize with its environment as an aesthetic contribution to its community. There are several key elements of design that create exemplary sustainable architecture:

- **Natural lighting.** The strategic use of natural lighting maximizes the energy efficiency of a facility and can reduce energy consumption costs. The first challenge is to incorporate natural light and windows in a design that opens up a tight, well-insulated building envelope without wasting energy. Natural lighting has a practical and emotional application. Humans don't want to live and work in hermetically sealed environments. We want windows and natural light, so the design team has to open up this sealed structure just enough to interact with the environment without jeopardizing its energy efficiency and the structural integrity of the building's envelope.

Figure 7.

- **Physical orientation.** The direction in which a building sits on a site directly affects its energy efficiency, environmental impact, and visual appeal. As an example, in regions with significant solar exposure, orienting buildings along an east-west axis responds to the diurnal path of the sun and makes optimal use of day light to reduce artificial light costs. Sun shading elements can add a layer of architectural detail to the exterior facades with the benefit of obstructing direct penetration of the sun's rays. These allow for the natural light to be bounced off of the shading devices into the interior volumes of the structure, thus diffusing the light and reducing much of the glare that can adversely affect the viewing of monitors and projection equipment. The east/west orientation also reduces the impact of solar radiation heating the building volume.

Figure 8.

- **Building materials.** Building materials—such as recyclable steel, low volatile organic compound (VOC) adhesives and carpeting, as well as roofing systems with a white reflective finish—directly impact a building's sustainability. There are also a variety of insulated metal wall panel systems that are highly recyclable and that can afford the facility a distinctive color and texture palette, and which come in a range of impact-rated systems that would be applicable for critical facilities.

Figure 9.

- **Color.** The visual power of color emphasizes the "tectonics" of a building and evokes emotional and physical reactions. The individual perception of color ranges vastly. There has been research that has focused on the mood-modifying behavioral aspects of color. Color can reduce stress, increase energy, and enhance productivity in professional working environments.

Figure 10.

- **Energy-efficient equipment/features.** Two design issues—energy load reduction and equipment specifications—play key roles in the creation of an energy-efficient building. Through design and prudent selection of active energy-consuming systems, such as HVAC and lighting, the design team can affect total performance by reducing energy consumption by up to half of a typical office facility. Interior lighting in office buildings tends to be the single largest element of energy consumption, accounting for about 30 percent of total energy costs, followed by heating/cooling costs.

Any good architectural team's approach to environmentally responsive architecture is explained simply by listening to the clients' need and implementing collective ideas into the preliminary planning and the final design and construction of a facility. This is achieved by following the "Recipe for Success:"

1. **Functional design:** Once the spatial needs are accurately documented, prepare a preliminary schematic design that serves the routine functions of the users and locates common areas effectively throughout the facility.

2. **Minimize energy consumption:** Address passive energy conservation in the design by utilizing shading devices, day lighting, insulation, high-efficiency heating and cooling systems, energy-saving lighting, appliances, and equipment.

3. **Physical orientation:** Orienting the structures with the long axis running east-west can affect energy efficiency, environmental materials, and visual appeal. If the site doesn't permit an east-west axis, then shield the building from direct sun through shading devices, creative landscaping, and take advantage of day lighting through the use of insulated translucent materials.

4. **Building insulation:** Insulate the building and specify the skin of the building according to the needs of the walls. Use insulated CMU or metal wall panel sys-

tems. If utilizing the insulated metal wall panel systems, specify those that have been tested for large missile impact conditions, as previously recommended.

5. **Promoting the design and construction of a healthy facility:** Minimize the use of toxic materials by specifying low- or no-odor finishing systems, formaldehyde-free materials, low/no-toxicity adhesives, and that no toxic cleaning solutions are required for maintenance. If the project is in a region with high humidity, detail the building envelope to ensure proper vapor barrier placement to minimize the possibility of humidity infiltration to maintain the appropriate level of indoor air quality to forbid the potential of mildew growth.

6. **Maintain indoor air quality:** Building commissioning procedures should be implemented to assure that indoor air quality is not only achieved at the time the facility becomes operational, but is subsequently maintained. All facilities should be designed to assure positive building pressure. When a building is over-vented and under negative indoor air pressurization, a variety of operational issues can result, such as increased heating and cooling loads/costs, and the potential of moisture intrusion, resulting in poor indoor air quality.

7. **Laminated/insulated glass:** Be aggressive in specifying reflective laminated and insulated glazing products. Additionally, evaluate the number of windows, and only include necessary windows. Encourage day lighting through the use of clerestories and light monitors on the roof to introduce natural light into the center core spaces within a structure that is not afforded an exterior wall for windows.

8. **Reduce power consumption proactively:** Specify T-8 fluorescent light tubes, motion-activated light switches, and energy-saving ballasts. These types of lights are the most cost-effective and return the initial investment very quickly with far less up-front costs. Also, plan the bulk of automated functions during non-peak times, according to the power supplier.

9. **Computerizing HVAC systems can be beneficial:** Tweak systems by specifying variable speed drives for the chiller pumps. Zoning the facility is a smart approach to reduce the overall heating and cooling loads in facilities that are operational on a 24-hour-per-day basis. This will allow for non-occupied areas to receive less heating or cooling when the personnel assigned to these areas are not on shift, thus reducing the operational costs for the lifecycle of the facility.

10. **Minimize or eliminate negative environmental impacts:** Plan the structure to minimize the use of expensive steel, wood, and paper to keep transportation costs and fuel consumption low. Encourage the use of recycled materials. Additionally, encourage the use of materials that have low maintenance requirements and long life spans.

11. **Stormwater recycling:** Plan the site and introduce pervious surfaces to allow stormwater recycling.

12. **Economical structural systems:** Specify and plan around economical structural systems.

13. **Recycled and recyclable building materials:** Encourage the use of recycled and recyclable building materials.

Although certain types of projects more readily lend themselves to sustainable architecture, the opportunity exists to apply environmentally responsive strategies to all types. Sustainable architecture reflects our modern-day priorities as stewards of the environment and our understanding of how humans, physical structures, and nature interact on a dynamic basis. Design professionals will continue to embrace and advance sustainable architecture in public sector, private, and commercial construction. As new discoveries are made, sustainable architecture continues to evolve under the basic tenet that there is no silver bullet that is going to solve everything. It's the quest to learn about the synergy between man-made and natural elements that makes sustainable architecture a reality.

Summary

An appropriate technical and conceptual approach to communications center design is based upon the recognition that there are very specialized spaces associated with this type of project, and a high level of expertise is needed to design a facility with these components. This facility type requires specialized systems such as security, communications, radio transmissions, audio/visual, specialized fire protection, HVAC, and electrical systems, as well as redundant back-up capacity for all of these components to function when primary sources of power, data, heating, and cooling are not functioning. These systems require conformance to a variety of specialized code requirements and guidelines.

All of these systems need to be successfully integrated into the building architecture by establishing the specific equipment and systems needs of all end users to be housed in the facility. This is accomplished by the design team of architects and engineers conducting detailed interviews with the key representatives from all departments to understand their specific job requirements and facility needs. The team, which includes specialized systems consultants, then works directly with the users to design a total systems package, including equipment, cabling distribution, and devices that not only meets the known current and future needs of the client, but also designs into the facility the noted components to afford the structure the appropriate capability to evolve as the user's needs and specialized equipment change.

The architects then provide space for the installation of these components. Mechanical and electrical engineers will provide cooling systems, grounding, electrical, and data power backup for optimal operation of the communication, security, and audio visual systems. Once the design is complete for all integrated systems, cost estimating is done, and prioritization of components are identified and completed.

IAN REEVES, AIA, serves as the president of Architects Design Group. His areas of specialized experience include 911 Communications, Law Enforcement, Public Safety Operations and Training, and Emergency Operations Center projects on a national basis. Mr. Reeves received his Master in Architecture degree from the University of Florida, and has spent his entire career with ADG, focusing his efforts on the advancement of public safety architecture. He is a leader in the industry of programming, master planning, funding strategies, and design of numerous successful projects. Mr. Reeves also is an annual speaker at the "Planning, Funding and Obtaining New Public Safety Facilities" seminar co-sponsored by the Center for Public Safety and Architects Design Group.

Made in the USA
Charleston, SC
29 September 2012